MONOGRAPHS ON STATISTICS AND APPLIED PROBABILITY

General Editors

1 Stochastic Population Models in Ecology and Epidemiology *M.S. Bartlett* (1960)
2 Queues *D.R. Cox and W.L. Smith* (1961)
3 Monte Carlo Methods *J.M. Hammersley and D.C. Handscomb* (1964)
4 The Statistical Analysis of Series of Events *D.R. Cox and P.A.W. Lewis* (1966)
5 Population Genetics *W.J. Ewens* (1969)
6 Probability, Statistics and Time *M.S. Bartlett* (1975)
7 Statistical Inference *S.D. Silvey* (1975)
8 The Analysis of Contingency Tables *B.S. Everitt* (1977)
9 Multivariate Analysis in Behavioural Research *A.E. Maxwell* (1977)
10 Stochastic Abundance Models *S. Engen* (1978)
11 Some Basic Theory for Statistical Inference *E.J.G. Pitman* (1979)
12 Point Processes *D.R. Cox and V. Isham* (1980)
13 Identification of Outliers *D.M. Hawkins* (1980)
14 Optimal Design *S.D. Silvey* (1980)
15 Finite Mixture Distributions *B.S. Everitt and D.J. Hand* (1981)
16 Classification *A.D. Gordon* (1981)
17 Distribution-free Statistical Methods, 2nd edition *J.S. Maritz* (1995)
18 Residuals and Influence in Regression *R.D. Cook and S. Weisberg* (1982)
19 Applications of Queueing Theory, 2nd edition *G.F. Newell* (1982)
20 Risk Theory, 3rd edition *R.E. Beard, T. Pentikainen and E. Pesonen* (1984)
21 Analysis of Survival Data *D.R. Cox and D. Oakes* (1984)
22 An Introduction to Latent Variable Models *B.S. Everitt* (1984)
23 Bandit Problems *D.A. Berry and B. Fristedt* (1985)
24 Stochastic Modelling and Control *M.H.A. Davis and R. Vinter* (1985)
25 The Statistical Analysis of Compositional Data *J. Aitchison* (1986)
26 Density Estimation for Statistics and Data Analysis *B.W. Silverman* (1986)
27 Regression Analysis with Applications *G.B. Wetherill* (1986)
28 Sequential Methods in Statistics, 3rd edition *G.B. Wetherill and K.D. Glazebrook* (1986)

29 Tensor Methods in Statistics *P. McCullagh* (1987)
30 Transformation and Weighting in Regression *R.J. Carroll and D. Ruppert* (1988)
31 Asymptotic Techniques for Use in Statistics *O.E. Barndorff-Nielsen and D.R. Cox* (1989)
32 Analysis of Binary Data, 2nd edition *D.R. Cox and E.J. Snell* (1989)
33 Analysis of Infectious Disease Data *N.G. Becker* (1989)
34 Design and Analysis of Cross-Over Trials *B. Jones and M.G. Kenward* (1989)
35 Empirical Bayes Methods, 2nd edition *J.S. Maritz and T. Lwin* (1989)
36 Symmetric Multivariate and Related Distributions *K.-T. Fang S. Kotz and K.W. Ng* (1990)
37 Generalized Linear Models, 2nd edition *P. McCullagh and J.A. Nelder* (1989)
38 Cyclic and Computer Generated Designs, 2nd edition *J.A. John and E.R. Williams* (1995)
39 Analog Estimation Methods in Econometrics *C.F. Manski* (1988)
40 Subset Selection in Regression *A.J. Miller* (1990)
41 Analysis of Repeated Measures *M.J. Crowder and D.J. Hand* (1990)
42 Statistical Reasoning with Imprecise Probabilities *P. Walley* (1991)
43 Generalized Additive Models *T.J. Hastie and R.J. Tibshirani* (1990)
44 Inspection Errors for Attributes in Quality Control *N.L. Johnson, S. Kotz and X. Wu* (1991)
45 The Analysis of Contingency Tables, 2nd edition *B.S. Everitt* (1992)
46 The Analysis of Quantal Response Data *B.J.T. Morgan* (1993)
47 Longitudinal Data with Serial Correlation: A State-space Approach *R.H. Jones* (1993)
48 Differential Geometry and Statistics *M.K. Murray and J.W. Rice* (1993)
49 Markov Models and Optimization *M.H.A. Davis* (1993)
50 Networks and Chaos - Statistical and Probabilistic Aspects *O.E. Barndorff-Nielsen, J.L. Jensen and W.S. Kendall* (1993)
51 Number-theoretic Methods in Statistics *K.-T. Fang and Y. Wang* (1994)
52 Inference and Asymptotics *O.E. Barndorff-Nielsen and D.R. Cox* (1994)
53 Practical Risk Theory for Actuaries *C.D. Daykin, T. Pentikäinen and M. Pesonen* (1994)
54 Biplots *J.C. Gower and D.J. Hand* (1996)
55 Predictive Inference: An Introduction *S. Geisser* (1993)
56 Model-Free Curve Estimation *M.E. Tarter and M.D. Lock* (1993)
57 An Introduction to the Bootstrap *B. Efron and R.J. Tibshirani* (1993)

58 Nonparametric Regression and Generalized Linear Models
P.J. Green and B.W. Silverman (1994)
59 Multidimensional Scaling *T.F. Cox and M.A.A. Cox* (1994)
60 Kernel Smoothing *M.P. Wand and M.C. Jones* (1995)
61 Statistics for Long Memory Processes *J. Beran* (1995)
62 Nonlinear Models for Repeated Measurement Data *M. Davidian and D.M. Giltinan* (1995)
63 Measurement Error in Nonlinear Models *R.J. Carroll, D. Ruppert and L.A. Stefanski* (1995)
64 Analyzing and Modeling Rank Data *J.I Marden* (1995)
65 Time Series Models:In econometrics, finance and other fields
D.R. Cox, D.V.Hinkley and O.E. Barndorff-Nielsen (1996)

(Full details concerning this series are available from the Publishers).

Time Series Models

In econometrics, finance and other fields

Edited by

D.R.Cox

Honorary Fellow
Nuffield College
Oxford, UK

D.V. Hinkley

Professor of Statistics
University of California
Santa Barbara, USA

and

O.E. Barndorff-Nielsen

Professor of Theoretical Statistics
Aarhus University
Denmark

CRC Press
Taylor & Francis Group
Boca Raton London New York

CRC Press is an imprint of the
Taylor & Francis Group, an **informa** business
A CHAPMAN & HALL BOOK

CRC Press
Taylor & Francis Group
6000 Broken Sound Parkway NW, Suite 300
Boca Raton, FL 33487-2742

First issued in paperback 2019

ISBN-13: 978-0-412-72930-0 (hbk)
ISBN-13: 978-0-367-40132-0 (pbk)

A Catalogue record for this book is available from the British Library

Visit the Taylor & Francis Web site at
http://www.taylorandfrancis.com

and the CRC Press Web site at
http://www.crcpress.com

Contents

Contributors

Neil G. Shephard, Nuffield College, Oxford, OX1 1NF, UK.

Søren Johansen, Institute of Mathematical Statistics, University of Copenhagen Ø, Universitetparken 5, 2100 Copenhagen, Denmark.

David F. Hendry, Nuffield College, Oxford, OX1 1NF, UK.

Michael P. Clements, Department of Economics, University of Warwick, Coventry, UK.

Nan Laird, Department of Biostatistics, Harvard School of Public Health, 677 Huntington Avenue, Boston, MA 02115-6096, USA.

Bjarne A. Jensen, Institute of Finance, Copenhagen Business School, Rosenørns Alle 31, 1970 Frederiksberg C, Denmark.

Jørgen A. Nielsen, Department of Theoretical Statistics and Operations Research, Institute of Mathematics, Aarhus University, Ny Muukegade, 8000 Aarhus C, Denmark.

Preface

This volume consists of the revised versions of the main papers given at the second Séminaire Européen de Statistique on 'Likelihood, Time Series, with Econometrics and Other Applications', held at Nuffield College, Oxford from 13–17 December 1994. The aim of the Séminaire Européen de Statistique is to provide talented young researchers with an opportunity to get quickly to the forefront of knowledge and research in areas of current major focus. Accordingly, as in the book based on the first seminar in the series, '*Networks and Chaos – Statistical and Probabilistic Aspects*', the papers in this volume have a tutorial character. In the present Séminaire about 35 young statisticians from ten European countries participated. Nearly all participants gave short presentations about their recent work; these, while of high quality, are not reproduced here.

The paper by N.G. Shephard reviews and extends work on a class of nonlinear time series models widely used in econometrics and of potential interest in other fields. S. Johansen gives a widely accessible account of cointegration, an important notion in the interpretation of multivariate nonstationary time series. M.P. Clements and D.F. Hendry give a general discussion of the statistics of forecasting errors. These three papers draw their motivation directly from econometrics. By contrast, N. Laird discusses methods developed in a biostatistical context for the analysis of short time series. Finally, B.A. Jensen and J.A. Nielsen develop from first principles the mathematical basis of option pricing and other topics in the mathematical theory of finance.

The second Séminaire Européen de Statistique was organized by O.E. Barndorff-Nielsen, Aarhus University; D.R. Cox and D. V. Hinkley, Oxford; W.S. Kendall, University of Warwick; N.G. Shephard, Nuffield College, Oxford; and C.N. Laws, Oxford. Future Séminaires Européens de Statistique are planned; the third will take place in Toulouse in 1996, on stochastic geometry.

The second Séminaire Européen de Statistique was supported by the

European Communities under an HCM Euroconferences contract, for which we express grateful appreciation.

On behalf of the Organizers
O.E. Barndorff-Nielsen, D.R. Cox and D.V. Hinkley
Aarhus and Oxford

CHAPTER 1

Statistical aspects of ARCH and stochastic volatility

Neil Shephard

1.1 Introduction

Research into time series models of changing variance and covariance, which I will collectively call **volatility models**, has exploded in the last ten years. This activity has been driven by two major factors. First, out of the growing realization that much of modern theoretical finance is related to volatility has emerged the need to develop empirically reasonable models to test, apply and deepen this theoretical work. Second, volatility models provide an excellent testing ground for the development of new nonlinear and non-Gaussian time series techniques.

There is a large literature on volatility models, so this chapter cannot be exhaustive. I hope rather to discuss some of the most important ideas, focusing on the simplest forms of the techniques and models used in the literature, referring the reader elsewhere for generalizations and regularity conditions. To start, I will consider two motivations for volatility models: empirical stylized facts and the pricing of contingent assets. In section 1.4 I will look at multivariate models, which play an important role in analysing the returns on a portfolio.

1.1.1 Empirical stylized facts

In most of this chapter I will work with two sets of financial time series. The first is a bivariate daily exchange rate series of the Japanese yen and the German Deutsche Mark measured against the pound sterling, which runs from 1 January 1986 to 12 April 1994, yielding 2160 observations. The second consists of the bivariate daily FTSE 100 and Nikkei 500 indexes, which are market indexes for the London and Tokyo equity markets. These series run from 2 April 1986 to 6 May 1994, yielding 2113 daily observations.

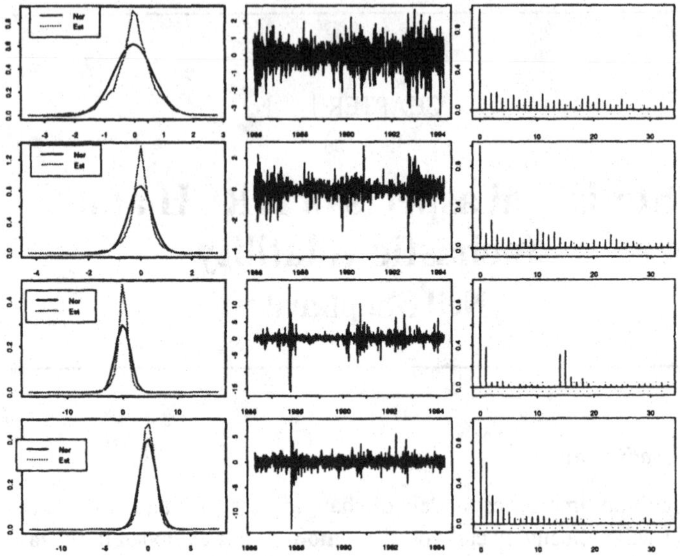

Figure 1.1 *Summaries of the daily returns on four financial assets. From top to bottom: yen, DM, Nikkei 500, FTSE 100. Summaries are: nonparametric density estimate and normal approximation, time series of returns and correlogram of the corresponding squares.*

Throughout I will work with the compounded return on the series $y_t = 100\log(x_t/x_{t-1})$ where x_t is the value of the underlying asset. Figure 1.1 displays some summaries of these two series. It gives a density estimate (using default S-Plus options) of the unconditional distribution of y_t together with the corresponding normal approximation. This suggests that y_t is heavy-tailed. This is confirmed by Table 1.1, which reports an estimate of the standardized fourth moments. In all but the Japanese case they are extremely large.

There is little evidence of any obvious forms of non-symmetry in the unconditional density. A correlogram of y_t shows little activity and so is not given in this figure; Figure 1.1 graphs the raw time series of y_t. Informally this picture suggests that there are periods of volatility clustering: days of large movements are followed by days with the same characteristics. This is confirmed by the use of a correlogram on y_t^2, and the corresponding Box-Ljung statistic reported in Table 1.1, which shows significant correlations which exist at quite extended lag lengths. This suggests that y_t^2 may follow a process close to an ARMA(1,1), for simple AR processes cannot easily combine the persistence in shocks with the

Table 1.1 *Summary statistics for the daily returns in Figure 1.1. BL denotes the Box–Ljung statistic, computed using the squares, with 30 lags. It should be around 30 if there is no serial dependence in the squares. K denotes the standardized fourth moment of the y_t. K should be around 3 under conditions of normality*

	BL	*K*
Japanese yen	563	5.29
German Deutsche Mark	638	36.2
Nikkei 500 index	828	36.2
FTSE 100 index	1375	28.1

low correlation. Finally, there is some evidence that the exchange rates and equity markets each share periods of high volatility, big movements in one currency being matched by large changes in another. This suggests that multivariate models will be important.

Reasons for changing volatility

It would be convenient to have an explanation for changing levels of volatility. One approach would be to assume that price changes occur as the result of a random number of intra-daily price movements, responding to information arrivals. Hence $y_t = \sum_{i=1}^{n_t} x_{it}$, where x_{it} are independently and identically distributed (i.i.d.) and n_t is some Poisson process. This type of model has a long history, going back to the work of Clark (1973). In this paper the n_t is assumed to be independent over time, which means y_t would inherit this characteristic. It is a trivial matter to allow n_t to be time-dependent, which would lead to volatility clustering, although the resulting econometrics becomes rather involved (see Tauchen and Pitts, 1983).

The more interesting literature in econometric terms is that which ties this information arrival interpretation into a model which also explains volume. The joint models of volume (see also Engle and Russell, 1994) and volatility are the focus of Gallant, Hsieh and Tauchen (1991), who use a reduced-form model, and of Andersen (1995). This is an interesting, but underdeveloped, area.

1.1.2 Pricing contingent assets

Suppose the value of some underlying security, written S, follows a geometric diffusion $dS = \mu S dt + \sigma S dz$, so that $d\log S = \left(\mu - \frac{\sigma^2}{2}\right) dt + \sigma dz$. It is possible to define an asset, c, which is a function of the underlying share price S. Economists call such assets **contingent** or **derivative**. Good introductions to the literature on this topic are given in Ingersoll (1987) and Hull (1993). A primary example of a contingent asset is an **option**, which allows the option owner the ability, but not the obligation, to trade the underlying asset at a given price in the future. The best-known example of this is the European call option whose owner can buy the underlying asset at the fixed price K, at the expiry date $T+v$. An example of K is where it equals $S(T)$, today's price; where the dependence of S on time is now shown explicitly. This special case is called an **at-the-money** option. The value of the general European call option at expiration will be

$$c(T+v) = \max\{S(T+v) - K, 0\}. \tag{1.1}$$

While equation (1.1) expresses the value of the option at time $T+v$, the option will be purchased at time T, so its purchase value has yet to be determined. A simple approach would be to compute the discounted expected value of the option,

$$\exp(-rv) E_{S(T+v)|S(T)} \max\{S(T+v) - K, 0\},$$

where r is a riskless interest rate. However, this neglects the fact that traders expect higher returns on risky assets than on riskless assets, a point which will recur in section 1.4. Hence the market will not typically value assets by their expected value. This suggests the introduction of a utility function into the pricing of options, allowing dealers to trade expected gain against risk.

It turns out that the added complexity of a utility function can be avoided by using some properties of diffusions and by assuming continuous and costless trading. This can be seen by constructing a portfolio worth π made up of owning θ of the underlying shares and by borrowing a single contingent asset c. Then the value of the portfolio evolves as

$$\begin{aligned} d\pi &= \theta dS - dc \\ &= \theta(\mu S dt + \sigma S dz) - (c_s \mu S + c_t + \tfrac{1}{2} c_{ss} \sigma^2 S^2) dt - c_s \sigma S dz \\ &= (\theta - c_s)(\mu S dt + \sigma S dz) - (c_t + \tfrac{1}{2} c_{ss} \sigma^2 S^2) dt, \end{aligned}$$

by using Itô's lemma, where $c_t = \partial c/\partial t$ and $c_s = \partial c/\partial S$. The investor, by selecting $\theta = c_s$ at each time period, can ensure $d\pi$ is instantaneously riskless by eliminating any dependence on the random dz. This result, of making $d\pi$ a deterministic function of time, is due to Black and Scholes (1973). As time passes, the portfolio will have continually to adjust to maintain risklessness – hence the need for continuous costless trading.

As this portfolio is riskless, its return must be the riskless interest rate

r, for otherwise traders will take this arbitrage opportunity and make instant riskless profits. The riskless interest rate can be taken as the return on a very short-duration government bond. Consequently the riskless portfolio follows

$$\begin{aligned} d\pi &= r\pi dt = r(c_s S - c)dt, \text{ as } \pi = c_s S - c \text{ to achieve risklessness} \\ &= -(c_t + \tfrac{1}{2}c_{ss}\sigma^2 S^2)dt, \end{aligned}$$

implying that the contingent asset follows the stochastic differential equation

$$c_t + \frac{1}{2}c_{ss}\sigma^2 S^2 + rc_s S = rc, \quad \text{with end condition } c = \max(S - K, 0).$$

This equation is remarkably simple. In particular, it does not depend on μ or the risk preference of the traders. Hence we can evaluate it as if the world was risk-neutral, in which case we can assume that the share price follows a new diffusion, with mean rS^* and variance $\sigma^2 S^{*2}$: $dS^* = rS^*\, dt + \sigma S^* dz$. This is the **risk-neutral** process; see Hull (1993, pp. 221–222). Using the log-normality of the diffusion we have

$$\log S^*(T+v) | \log S(T) \sim N\{\log S(T) + (r - \sigma^2/2)v, \sigma^2 v\}.$$

Straightforward log-normal results give us the Black–Scholes valuation of the option v periods ahead, using an instantaneous variance of σ^2, of

$$bs_v(\sigma^2) = \exp(-rv)E\left[\max\{S^*(T+v) - K, 0\} | S(T)\right]$$

which is

$$bs_v(\sigma^2) = S(T)\Phi(d) - K\exp(-rv)\Phi(d - \sigma\sqrt{v}), \qquad (1.2)$$

where

$$d = \frac{\log\{S(T)/K\} + (r + \sigma^2/2)v}{\sigma\sqrt{v}}. \qquad (1.3)$$

Note that v and K are given by institutional norms, $S(T)$ and r are observed, leaving only σ^2 as unknown. In a real sense, option prices are valuing volatility. As with much of finance, it is the volatility which plays the crucial role, rather than the mean effect.

Empirically there are two straightforward ways of using (1.2). The first is to estimate σ^2 and then work out the resulting option price. The second is to use the observed option prices to back out a value for σ^2. This second method is called an **implied volatility estimate**; see Xu and Taylor (1994) for a modern treatment of this.

A difficulty with all of this analysis is the basic underlying assumption of the process, that stock returns follow a geometric diffusion. Figure 1.1 indicates that this is a poor assumption, in turn suggesting that (1.2) may give a poor rule on which to base option pricing. This realization has prompted theoretical work into option pricing theory under various changing volatility regimes. The leading paper in this field is by Hull and White (1987). I will return to this later.

1.1.3 Classifying models of changing volatility

There are numerous models of changing variance and covariance. A useful conceptual division of the models, following Cox (1981), is into **observation-driven** and **parameter-driven** models. For convenience I will discuss these two approaches within the confines of a tightly defined parametric framework which allows

$$y_t \mid z_t \sim N(\mu_t, \sigma_t^2).$$

For compactness of exposition μ_t will often be set to zero as I do not intend to focus on that feature of the model. Observation-driven models put z_t as a function of lagged values of y_t. The simplest example of this was introduced by Engle (1982) in his paper on autoregressive conditional heteroscedasticity (ARCH). This allows the variance to be a linear function of the squares of past observations

$$\sigma_t^2 = \alpha_0 + \alpha_1 y_{t-1}^2 + \ldots + \alpha_p y_{t-p}^2,$$

and so the model becomes one for the one-step-ahead forecast density:

$$y_t \mid Y_{t-1} \sim N(0, \sigma_t^2),$$

where Y_{t-1} is the set of observations up to time $t-1$. This allows today's variance to depend on the variability of recent observations.

Models built out of explicit one-step-ahead forecast densities are compelling for at least three reasons. First, from a statistical viewpoint, combining these densities delivers the likelihood via a prediction decomposition. This means estimation and testing are straightforward, at least in principle. Second, and more importantly from an economic viewpoint, finance theory is often specified using one-step-ahead moments, although it is defined with respect to the economic agents' information set not the econometricians'. The third reason for using observation-driven models is that they parallel the very successful autoregressive and moving average models which are used so widely for models of changing means. Consequently some of the techniques which have been constructed for these models can be used for the new models. ARCH type models have attracted a large amount of attention in the econometrics literature. Surveys of this work are given in the papers by Bollerslev, Chou and Kroner (1992), Bollerslev, Engle and

Nelson (1995), Bera and Higgins (1995) and Diebold and Lopez (1995). Finally, Engle (1995) is an extensive reprint collection of ARCH papers.

Parameter-driven or state-space models allow z_t to be a function of some unobserved or latent component. A simple example of this is the log-normal stochastic variance or volatility (SV) model, due to Taylor (1986):

$$y_t \mid h_t \sim N\left\{0, \exp(h_t)\right\}, \quad h_{t+1} = \gamma_0 + \gamma_1 h_t + \eta_t, \quad \eta_t \sim NID(0, \sigma_\eta^2),$$

where *NID* denotes normally and independently distributed. Here the log-volatility h_t is unobserved (at least by the econometrician) but can be estimated using the observations. These models parallel the Gaussian state-space models of means dealt with by Kalman (1960) and highlighted by Harrison and Stevens (1976) and West and Harrison (1989). In econometrics this type of models is associated with the work of Harvey (1989).

Unfortunately, unlike the models of the mean which fit into the Gaussian state-space form, almost all parameter-driven volatility models lack analytic one-step-ahead forecast densities $y_t \mid Y_{t-1}$. As a result, in order to deal with these models, either approximations have to be made or numerically intensive methods used. There seems to be only one constrained exception to this: stochastic volatility models which possess analytic filtering algorithms. Shephard (1994a) suggests setting h_{t+1} to be a random walk with $\exp(\eta_t)$ using a highly contrived scaled beta distribution, following some earlier work on some different non-Gaussian models by Smith and Miller (1986) and Harvey and Fernandes (1989). This delivers a one-step-ahead prediction distribution which has some similarities to the ARCH model. It has been generalized to the multivariate case by Uhlig (1992), who uses it to allow the covariance matrix of the innovations of a vector autoregression to change in a highly parsimonious way. Unfortunately, it does not seem possible to move away from h_{t+1} being a random walk without losing conjugacy. This inflexibility is worrisome and suggests this approach may be a dead end.

Although SV models are harder to handle statistically than the corresponding observation-driven models, there are some good reasons for still investigating them. We will see that their properties are easier to find, understand, manipulate and generalize to the multivariate case. They also have simpler analogous continuous-time representations, which is important given that much of modern finance employs diffusions. An example of this is the work by Hull and White (1987) which uses a log-normal SV model, replacing the discrete-time AR(1) for h_{t+1} with an Ornstein–Uhlenbeck process. A survey of some of the early work on SV models is given in Taylor (1994).

Basic statistical background

To understand the properties of volatility models it is important to have careful definitions of some of the most basic time series concepts for, unusually in time series modelling, small differences in these definitions can have substantial impact. The most commonly used will be:

- White noise (WN). This means $E(y_t) = \mu$, $\text{var}(y_t) = \sigma^2$ and $\text{cov}(y_t, y_{t+s}) = 0$, for all $s \neq 0$. Often μ will be taken to be zero. These unconditional moment conditions are sometimes strengthened to include y_t being independent, rather than uncorrelated, over time. This will be called strong WN, a special case of which is i.i.d.
- Martingale difference (MD). A related concept, y_t being MD stipulates that $E|y_t| < \infty$ and that $E(y_t|Y_{t-1}) = 0$. All MDs have zero means and are uncorrelated over time. If the unconditional variance of the MD is constant over time, then the series is also white noise.
- Covariance stationarity. This generalizes WN to allow autocovariance of the form $\text{cov}(y_t, y_{t+s}) = \gamma(s)$ for all t : the degree of covariance among the observations depends only on the time gap between them. The notation $\text{corr}(y_t, y_{t+s}) = \rho(s) = \gamma(s)/\sigma^2$ denotes the autocorrelation function.
- Strict stationarity. For some models moments will not exist, even in cases where the corresponding unconditional distributions are perfectly well behaved. As a result strict stationarity, where $F(y_{t+h}, y_{t+h+1}, \ldots, y_{t+h+p}) = F(y_t, y_{t+1}, \ldots, y_{t+p})$ for all p and h, will play a particularly prominent role.

1.2 ARCH

The simplest linear ARCH model, ARCH(1), puts:

$$y_t = \varepsilon_t \sigma_t, \quad \sigma_t^2 = \alpha_0 + \alpha_1 y_{t-1}^2, \quad t = 1, \ldots, T, \tag{1.4}$$

where $\varepsilon_t \sim NID(0, 1)$. The parameter α_1 has to be non-negative to ensure that $\sigma_t^2 \geq 0$ for all t. Crucially $y_t \mid Y_{t-1} \sim N(0, \sigma_t^2)$, which means y_t is a MD and under strict stationarity has a symmetric unconditional density. To show it is zero-mean white noise, we need to find its variance. Clearly the model can be written as a non-Gaussian autoregression:

$$y_t^2 = \sigma_t^2 + (y_t^2 - \sigma_t^2) = \alpha_0 + \alpha_1 y_{t-1}^2 + v_t, \tag{1.5}$$

where $v_t = \sigma_t^2(\varepsilon_t^2 - 1)$ and the sign of y_t is randomized. As v_t is a martingale difference, then if $\alpha_1 \in [0, 1)$, $E(y_t^2) = \alpha_0/(1 - \alpha_1)$. After some effort, it can be seen that

$$E(y_t^4)/\{E(y_t^2)\}^2 = 3(1-\alpha_1^2)/(1-3\alpha_1^2),$$

if $3\alpha_1^2 < 1$, which when it exists is greater than 3. Under this tight condition, y_t^2 is covariance stationary, its autocorrelation function is $\rho_{y_t^2}(s) = \alpha_1^s$, and y_t has leptokurtosis (fat tails). Notice that $\rho_{y_t^2}(s) \geq 0$ for all s, a result which is common to all linear ARCH models.

These are interesting results. If $\alpha_1 < 1$, y_t is white noise while y_t^2 follows an autoregressive process, yielding volatility clustering. However, y_t^2 is not necessarily covariance stationary for its variance will be finite only if $3\alpha_1^2 < 1$.

The conflicting conditions for covariance stationarity for y_t and y_t^2 prompt the interesting question as to the condition needed on α_0 and α_1 to ensure strict stationarity for y_t. This can be found as a special case of the results of Nelson (1990a), who proved that α_1 had to satisfy $E\{\log(\alpha_1\varepsilon_t^2) < 0\}$, which in Gaussian models implies that $\alpha_1 < 3.5622$.

1.2.1 Estimation

At first sight it is tempting to use the autoregressive representation (1.5) to estimate the parameters of the model (this was used by Poterba and Summers, 1986). If v_t is white noise this can be carried out by least squares; in effect this estimate will be reported as the first spike of the correlogram for y_t^2. Although a best linear unbiased estimator, this estimate would be inefficient.

ARCH models, like all observation-driven models, are designed to allow the likelihood to be found easily. Using a prediction decomposition (and ignoring constants):

$$\begin{aligned}\log f(y_1,\ldots,y_T \mid y_0;\theta) &= \sum_{t=1}^{T} \log f(y_t \mid Y_{t-1};\theta) \\ &= -\frac{1}{2}\sum_{t=1}^{T}\log\sigma_t^2 - \frac{1}{2}\sum_{t=1}^{T} y_t^2/\sigma_t^2 \quad (1.6)\end{aligned}$$

where θ will denote the parameters which index the model, in this case $(\alpha_0, \alpha_1)'$.

Notice that this likelihood conditions on some prior observations (or in real problems, the first few observations). This is convenient since the analytic form for the unconditional distribution of an ARCH model is unknown. The consequence is that the likelihood does not impose $\alpha_1 < 1$.

Table 1.2 *Aspects of the distribution of the ML estimator of ARCH(1) and GARCH(1,1). RMSE denotes root mean square error. ARCH true values of* $\alpha_0 = 0.2$ *and* $\alpha_1 = 0.9$*. GARCH true values of* $\alpha_0 = \alpha_1 = 0.2$ *and* $\beta_1 = 0.7$*. Based on 1000 replications. Top table – ARCH, bottom table – GARCH*

ARCH

T	$E(\hat{\alpha}_1)$	RMSE$(\hat{\alpha}_1)$	$\Pr(\hat{\alpha}_1 \geq 1)$
100	0.85221	0.25742	0.266
250	0.88386	0.16355	0.239
500	0.89266	0.10659	0.152
1000	0.89804	0.08143	0.100

GARCH

T	$E(\hat{\alpha}_1 + \hat{\beta}_1)$	RMSE$(\hat{\alpha}_1 + \hat{\beta}_1)$	$\Pr(\hat{\alpha}_1 + \hat{\beta}_1 \geq 1)$
100	0.87869	0.14673	0.206
250	0.88680	0.10246	0.143
500	0.89680	0.06581	0.060
1000	0.89913	0.04893	0.019

It is possible to find the scores for the model:

$$\frac{\partial \log f}{\partial \theta} = \frac{1}{2} \sum_{t=1}^{T} \frac{\partial \sigma_t^2}{\partial \theta} \frac{1}{\sigma_t^2} \left(\frac{y_t^2}{\sigma_t^2} - 1 \right), \tag{1.7}$$

where $\partial \sigma_t^2 / \partial \theta = (1, y_{t-1}^2)'$. Typically, even for such a simple model, the likelihood tends to be rather flat unless T is quite large. This means that the resulting maximum likelihood (ML) estimates of α_0 and α_1 are quite imprecise. Table 1.2 gives an example of this, reporting the mean and root mean squared error for the ML estimate of α_1. Notice the substantial probability that the estimated model is not covariance stationary, even when $T = 1000$.

The asymptotic behaviour of the ML estimation of the ARCH model has been studied by Weiss (1986) who showed normality if y_t has a bounded fourth moment. Unfortunately this rules out most interesting ARCH models. More recently Lumsdaine (1991) and Lee and Hansen (1994) have relaxed this condition substantially. Further, both papers look at the consequences of the possible failure of the normality assumption on ε_t (see Bollerslev and Wooldridge, 1992). By relaxing this assumption they treat (1.6) as a quasi-likelihood, which still ensures consistent estimation, but requires the use of the robust sandwich variance estimator (for the calculation of the variance of the estimators, see White, 1982; 1994).

Lee and Hansen (1994) state the following two main sufficient conditions for the consistency of the quasi-likelihood estimator using (1.6):

1. $E(\varepsilon_t \mid Y_{t-1}) = 0, \quad E(\varepsilon_t^2 \mid Y_{t-1}) = 1;$
2. $E\{\log(\alpha_1 \varepsilon_t^2) \mid Y_{t-1}\} < 0.$

The first ensures that the ARCH model correctly specifies the first two moments, the second that y_t is strictly stationary. Asymptotic normality additionally requires that $E(\varepsilon_t^4 \mid Y_{t-1})$ is bounded and that $\alpha_0, \alpha_1 > 0$.

1.2.2 Non-normal conditional densities

The Gaussian assumption on ε_t is arbitrary, indicating that we should explore other distributions. Although ARCH models can display fat tails, the evidence of the very fat-tailed unconditional distributions found for financial data (Mandelbrot, 1963; Fama, 1965) suggests that it may be useful to use models based on distributions with fatter tails than the normal distribution. Obvious candidate distributions include the Student t, favoured by Bollerslev (1987), and the generalized error distribution, used by Nelson (1991); see Evans, Hastings and Peacock (1993) for details of this error distribution. Notice that in both cases it is important to define the new ε_t so that it has unit variance.

Finally, there has recently been considerable interest in the development of estimation procedures which either estimate semi-parametrically the density of ε_t (Engle and Gonzalez-Rivera, 1991) or adaptively estimate the parameters of ARCH models in the presence of a non-normal ε_t (Steigerwald, 1991; Linton 1993). These seem promising areas of research; however, given that parametric estimation of ARCH models requires such large data sets, their effectiveness for real data sets seems questionable.

1.2.3 Testing for ARCH

Using the score (1.7) and corresponding Hessian it is possible to construct a score test of the hypothesis that $\alpha_1 = 0$, i.e. there is no volatility clustering in the series. It turns out to be the natural analogue of the portmanteau score test for AR(1) or MA(1), but in the squares. A generalization to more complicated ARCH models results in the analogue of the Box–Pierce statistic, which uses serial correlation coefficients for the squares

$$r_j = \sum \left(y_t^2 - \overline{y^2}\right)\left(y_{t-j}^2 - \overline{y^2}\right) / \sum \left(y_t^2 - \overline{y^2}\right)^2$$

rather than for the levels. It is studied in Engle, Hendry and Trumble (1985).

More recently, in the econometric literature, some concern has been expressed about the fact that these types of test do not exploit the full information about the model. In particular, $\sigma_t^2 \geq 0$ and so $\alpha_0, \alpha_1 \geq 0$, so that tests of the null hypothesis and more complicated variants of it have to be one-sided. Papers which address this issue include Lee and King (1993) and Demos and Sentana (1994).

1.2.4 Forecasting

One of the aims of building time series models is to be able to forecast. In ARCH models attention focuses not on $E(y_{T+s} \mid Y_T)$ as this is zero, but rather on $E(y_{T+s}^2 \mid Y_T)$ or more usefully, in my opinion, the whole distribution of $y_{T+s} \mid Y_T$.

In ARCH models it is easy to evaluate the forecast moments of $E(y_{T+s} \mid Y_T)$ (see Engle and Bollerslev, 1986; Baillie and Bollerslev, 1992). In the ARCH(1) case

$$E(y_{T+s}^2 \mid Y_T) = \alpha_0(1 + \alpha_1 + \ldots + \alpha_1^{s-1}) + \alpha_1^s y_T^2.$$

In non-covariance stationary cases, such as when $\alpha_1 = 1$, this forecast continually trends upwards, going to infinity with s. This may be somewhat unsatisfactory for some purposes, although if there is not much persistence in the process a normal approximation based on $y_{T+s} \mid Y_T \sim N\{0, E(y_{T+s}^2 \mid Y_T)\}$ may not be too unsatisfactory. That is the conclusion of Baillie and Bollerslev (1992).

In more complicated models it seems sensible to have simple methods to estimate informatively and report the distribution of $y_{T+s} \mid Y_T$. This is studied, using simulation, by Geweke (1989), who repeatedly simulates (1.4) into the future M times, and summarizes the results. A useful graphical representation of the simulation results is the plot of various estimated quantiles of the distribution against s. The results of Koenker, Ng and Portnoy (1994) are useful in reducing the required amount of simulation through smoothing quantile techniques.

1.2.5 Extensions of ARCH

The basic univariate ARCH model has been extended in a number of directions, some dictated by economic insight, others by broadly statistical ideas. The most important of these is the extension to include moving average parts, namely the generalized ARCH (GARCH) model. Its simplest example is GARCH(1,1) which puts

$$y_t = \varepsilon_t \, \sigma_t, \quad \sigma_t^2 = \alpha_0 + \alpha_1 y_{t-1}^2 + \beta_1 \sigma_{t-1}^2.$$

This model is usually attributed to Bollerslev (1986), although it was formulated simultaneously by Taylor (1986). It has been tremendously successful in empirical work and is regarded as the benchmark model by many econometricans.

GARCH

The GARCH model can be written as a non-Gaussian linear ARMA model in the squares:

$$y_t^2 = \alpha_0 + \alpha_1 y_{t-1}^2 + \beta_1 \sigma_{t-1}^2 + v_t = \alpha_0 + (\alpha_1 + \beta_1) y_{t-1}^2 + v_t - \beta_1 v_{t-1}, \tag{1.8}$$

following (1.5). The original series y_t is covariance stationary if $\alpha_1 + \beta_1 < 1$. In practice the fourth moment of y_t will not usually exist (see the conditions needed in Bollerslev, 1986), but y_t will be strictly stationary if $E \log(\beta_1 + \alpha_1 \varepsilon_t^2) < 1$ and $\alpha_0 > 0$. Nelson (1990a) graphs the combinations of α_1 and β_1 that this allows: importantly, it does include $\alpha_1 + \beta_1 \leq 1$.

The case of $\alpha_1 + \beta_1 = 1$ has itself received considerable attention. It is called integrated GARCH (IGARCH) (see Bollerslev and Engle, 1993). We will see later that for many empirical studies $\alpha_1 + \beta_1$ is estimated to be close to one, indicating that volatility has quite persistent shocks.

In ARCH models the likelihood can be constructed by conditioning on initial observations. In the GARCH(1,1) model both σ_{t-1}^2 and y_{t-1}^2 are required. A standard approach to this problem is to use an initial stretch of 20 observations, say, to calculate σ_{21}^2 by using a simple global variance estimate and computing $\log f(y_{21}, \ldots, y_T | \sigma_{21}^2, y_{20}^2; \theta)$. This is somewhat unsatisfactory, although for large n the initial conditions will not have a substantial impact. Standard normal asymptotics have been proved so long as y_t is strictly stationary (see Lee and Hansen, 1994). Interestingly asymptotic normality does hold for the unit root case, $\alpha_1 + \beta_1 = 1$, unlike for the corresponding Gaussian AR models studied in, for example, Phillips and Durlauf (1986).

To glean some idea of the sampling behaviour of the ML estimator for this model, I repeat the ARCH(1) simulation experiment but now with $\alpha_0 = \alpha_1 = 0.2$ and $\beta_1 = 0.7$. Table 1.2 reports the properties of $\widehat{\alpha}_1 + \widehat{\beta}_1$, as this is the most meaningful parametrization. It inherits most of the properties we found for the ARCH(1) model. Again there is a substantial probability of estimating this persistence parameter as being greater than one.

This model can be generalized by allowing p lags of y_t^2 and q lags of σ_t^2 to enter σ_t^2. This GARCH(p, q) is also strictly stationary in the integrated case, an extension of the GARCH(1,1) case proved by Bougerol and Picard (1992). This suggests normal asymptotics for the ML estimator can also be used in this more complicated situation.

Log GARCH

To statisticians ARCH models may appear somewhat odd. After all $y_t^2 = \varepsilon_t^2\sigma_t^2$ is a scaled χ_1^2 or gamma variable. Usually when we model the changing mean of a gamma distribution, a log link is used in the generalized linear model (see, for example, McCullagh and Nelder, 1989, Chapter 8). Consequently for many readers a natural alternative to this model might be

$$y_t^2 = \varepsilon_t^2 \exp(h_t), \quad h_t = \gamma_0 + \gamma_1 \log y_{t-1}^2.$$

This suggestion has been made by Geweke (1986) but has attracted little support. A major reason for this is that y_t is often close to zero (or quite often exactly zero). In a rather different context Zeger and Qaqish (1988) have proposed a simple solution to this problem by replacing h_t by

$$h_t = \gamma_0 + \gamma_1 \log\{\max(y_{t-1}^2, c)\}, \quad c > 0.$$

The constant c is a nuisance parameter which can be estimated from the data.

Exponential GARCH

Although the log GARCH models have not had very much impact, another log-based model has, but for rather different reasons. Nelson (1991) introduced an exponential GARCH (EGARCH) model for h_t which in its simplest form is

$$h_t = \gamma_0 + \gamma_1 h_{t-1} + g(\varepsilon_{t-1}), \quad \text{where } g(x) = wx + \lambda(|\, x \,| - E \,|\, x \,|). \tag{1.9}$$

The $g(\cdot)$ function allows both the size and sign of its argument to influence its value. Consequently when $\varepsilon_{t-1} > 0, \partial h_t/\partial\varepsilon_{t-1} = w + \lambda$, while the derivative is $w - \lambda$ when $\varepsilon_{t-1} < 0$. As a result EGARCH responds non-symmetrically to shocks.

The Nelson (1991) paper is the first which models the conditional variance as a function of variables which are not solely squares of the observations. The asymmetry of information is potentially useful for it allows the variance to respond more rapidly to falls in a market than to corresponding rises. This is an important stylized fact for many assets (see Black, 1976; Schwert, 1989; Sentana, 1991; Campbell and Hentschel, 1992). The EGARCH model is used on UK stocks by Poon and Taylor (1992).

Although (1.9) looks somewhat complicated, its properties are quite easy to find. As $\varepsilon_{t-1} \sim$ i.i.d., so $g(\varepsilon_{t-1}) \sim$ i.i.d.. It also has zero mean and a constant variance (ε_t is uncorrelated with $\mid \varepsilon_t \mid - E \mid \varepsilon_t \mid$ due to the symmetry of ε_t). As a result, like (1.8), h_t is an autoregression and so is stationary if and only if $\mid \gamma_1 \mid < 1$. Notice that this allows $\rho_{y_t^2}(s)$ to be negative, unlike linear ARCH models. Hence EGARCH models can produce cycles in the autocorrelation function for the squares.

Like the ARCH model, EGARCH is built to allow the likelihood function to be easily evaluated. At present the limit theory for this model has not been rigorously examined, although it seems clear that asymptotic normality will be obtained if $\mid \gamma_1 \mid < 1$.

Decomposing IGARCH

GARCH can be extended to allow arbitrary numbers of lags on y_t^2 and σ_t^2 to enter the variance predictor. A difficulty with this approach is a lack of parsimony, due to the absence of structure in the model. Recently Engle and Lee (1992) have addressed this issue by parametrizing a GARCH model into permanent and transitory components, analogous to the Beveridge and Nelson (1981) decomposition for means. A simple example is

$$\begin{aligned} \sigma_t^2 &= \mu_t + \alpha_1(y_{t-1}^2 - \mu_t) + \beta_1(\sigma_{t-1}^2 - \mu_t) \\ \mu_t &= w + \mu_{t-1} + \phi(y_{t-1}^2 - \sigma_{t-1}^2). \end{aligned}$$

Here the intercept of the GARCH process, μ_t, changes over time. As $\Delta\mu_t - w$ is an MD, μ_t is a persistent process tracing the level of the volatility process while σ_t^2 deals with the temporary fluctuations.

It is possible to rewrite this model into its reduced form

$$\begin{aligned} \sigma_t^2 &= w(1 - \alpha_1 - \beta_1) + \{\alpha_1 + \phi(1 - \alpha_1 - \beta_1)\}y_{t-1}^2 - \alpha_1 y_{t-2}^2 \\ &+ \{1 + \beta_1 - \phi(1 - \alpha_1 - \beta_1)\}\sigma_{t-1}^2 - \beta_1\sigma_{t-2}^2, \end{aligned}$$

which is a constrained IGARCH(2,2) model.

Fractionally integrated ARCH

Volatility tends to change quite slowly, with the effects of shocks taking a considerable time to decay (see Ding, Granger and Engle, 1993). This indicates that it might be useful to exploit a fractionally integrated model. The nonlinear autoregressive representation of ARCH suggests starting with:

$$(1-L)^d y_t^2 = \alpha_0 + v_t, \quad v_t = \sigma_t^2(\varepsilon_t^2 - 1), \quad d \in (-0.5, 0.5),$$

as the simplest fractionally integrated ARCH (FIARCH) model. Rewritten, this gives, say

$$\sigma_t^2 = \alpha_0 + \{1-(1-L)^d\}y_t^2 = \alpha_0 + \alpha(L)y_{t-1}^2.$$

Here $\alpha(L)$ is a polynomial in L which decays hyperbolically in lag length, rather than geometrically. Generalizations of this model introduced by Baillie, Bollerslev and Mikkelsen (1995), straightforwardly transform the ARFIMA models developed by Granger and Joyeux (1980) and Hosking (1981) into long-memory models of variance.

Although these models are observation-driven and so it is possible to write down $f(y_t \mid Y_{t-1})$, Y_{t-1} now has to contain a large amount of relevant data due to the slow rate of decay in the influence of old observations. This is worrying, because the likelihood of ARCH models usually conditions on some Y_0, working with $f(y_1, \ldots, y_T \mid Y_0)$. I think that for these models the construction of Y_0 may be important, although Baillie, Bollerslev and Mikkelsen (1995) argue this is not the case.

Weak GARCH

In this chapter emphasis has been placed on parametric models, which in the ARCH case means models of one-step-ahead prediction densities. Recently there has been some interest in weakening these assumptions, for a variety of reasons. One approach, from Drost and Nijman (1993), is to introduce a class of 'weak' GARCH models which do not build σ_t^2 out of $E(y_t^2 \mid Y_{t-1})$, but instead work with a best linear projection in terms of $1, y_{t-1}, y_{t-2}, \ldots, y_{t-1}^2, \ldots, y_{t-p}^2$.

Weak GARCH has been a useful tool in the analysis of temporally aggregated ARCH processes (see Drost and Nijman, 1993; Nijman and Sentana, 1993) and the derivation of continuous-time ARCH models (Drost and Werker, 1993). However, inference for these models is not trivial for it relies upon equating sample autocorrelation functions with their population analogues. This type of estimator can be ill behaved if y_t^2 is not covariance stationary (a tight condition).

Unobserved ARCH

A number of authors, principally Diebold and Nerlove (1989), Harvey, Ruiz and Sentana (1992), Gourieroux, Monfort and Renault (1993) and King, Sentana and Wadhwani (1994) have studied ARCH models observed with error:

$$y_t = f_t + \eta_t, \quad f_t = \varepsilon_t \sigma_t, \quad \sigma_t^2 = \alpha_0 + \alpha_1 f_{t-1}^2, \tag{1.10}$$

where ε_t and η_t are mutually independent and are normally and independently distributed. Their variances are 1 and σ_η^2 respectively. Unlike the other ARCH-type models outlined above, (1.10) is not easy to estimate for it is not possible to deduce $f(y_t \mid Y_{t-1})$ analytically. This is because f_{t-1} is not known given Y_{t-1}. Hence, it makes sense to think of these models as parameter-driven and so classify them as stochastic volatility models.

Approaches to tackling the problem of inference for this model are spelt out by Harvey, Ruiz and Sentana (1992). They employ a Kalman filtering approach based on the state space:

$$\begin{aligned} y_t &= f_t + \eta_t, & \eta_t &\sim NID(0, \sigma_\eta^2), \\ f_t &= f_t, & f_t &\sim N(0, \alpha_0 + \alpha_1 f_{t-1}^2). \end{aligned}$$

It is possible to estimate f_t and η_t, using the unconditional distribution of f_t as the disturbance of the transition equation. The resulting filter gives a best linear estimator. However, it is inefficient because it ignores the dynamics.

An alternative approach is to use Y_t to estimate f_t and then use that estimate, $\widehat{f_t}$, to adapt the variance of f_{t+1} so that $f_{t+1} \mid Y_t \overset{\cdot}{\sim} N(0, +\alpha_0 + \alpha_1 \widehat{f_t}^2)$. This is the approach of Diebold and Nerlove (1989). The approximation can be improved by noting that

$$f_t^2 = \widehat{f_t}^2 + (f_t - \widehat{f_t})^2 + 2\widehat{f_t}(f_t - \widehat{f_t}).$$

Taking expectations of this, given Y_t, and using the approximation $\widehat{f_t} \simeq E(f_t | Y_t)$, yields

$$E(f_t^2 \mid Y_t) \simeq \widehat{f_t}^2 + p_t, \text{ where } p_t \simeq E\{(f_t - \widehat{f_t})^2 \mid Y_t\}.$$

This delivers the improved approximation $f_{t+1} \mid Y_t \overset{\cdot}{\sim} N\{0, \alpha_0 + \alpha_1(\widehat{f_t}^2 + p_t)\}$. As $\widehat{f_t}$ and p_t are in Y_t, if this were the true model, the resulting Kalman filter would be optimal; as it is, Harvey, Ruiz and Sentana (1992) use the phrase 'quasi-optimal' to describe their result. However, as it does not seem possible to prove any properties about this 'quasi-optimal' filter, perhaps a better name would be an 'approximate filter'.

A likelihood-based approach to this model is available via a Markov chain Monte Carlo (MCMC) method, since the model has a Markov random fields structure and so

$$f(f_t|f_{\backslash t}, y) \propto f(f_t|f_{t-1})f(f_{t+1}|f_t)f(y_t|f_t),$$

the notation $f_{\backslash t}$ meaning all elements of $f_1, \ldots, f_n$ except f_t. By continually simulating from $f_t|f_{\backslash t}, y$, for $t = 1, \ldots, n$, a Gibbs or Metropolis sampler can be constructed which converges to a sample from $f_1, \ldots, f_n|y$. These techniques will be spelt out in more detail in the next section.

Other ARCH specifications

There have been numerous alternative specifications for ARCH models. Some of the more influential include those based on:

- Absolute residuals. Suggested by Taylor (1986) and Schwert (1989), this puts:

$$\sigma_t = \alpha_0 + \alpha_1 \mid y_{t-1} \mid .$$

- Nonlinear. The NARCH model of Engle and Bollerslev (1986) and Higgins and Bera (1992) has the flavour of a Box–Cox generalization. It allows:

$$\sigma_t^2 = \alpha_0 + \alpha_1 \mid y_{t-1} \mid^{\gamma},$$

or a non-symmetric version:

$$\sigma_t^2 = \alpha_0 + \alpha_1 \mid y_{t-1} - k \mid^{\gamma} .$$

- Partially nonparametric model. Early works by Pagan and Schwert (1990) and Gourieroux and Monfort (1992) have tried to let the functional form of σ_t^2, as a response to y_{t-1}, be determined empirically. A simple approach is given in Engle and Ng (1993) who use a linear spline for σ_t^2:

$$\begin{aligned}\sigma_t^2 = \alpha_0 \quad &+ \sum_{j=0}^{m+} \alpha_1^{+j} I(y_{t-1} - \tau_j > 0)(y_{t-1} - \tau_j) \\ &+ \sum_{j=0}^{m^-} \alpha_1^{-j} I(y_{t-1} - \tau_j < 0)(y_{t-1} - \tau_{-j}),\end{aligned}$$

where $I(\cdot)$ are indicator functions and $(\tau_{-m}, \ldots, \tau_m)$ is an ordered set of knots typically set as $\tau_j = j\sqrt{\text{var}(y_t)}$ with $\tau_0 = 0$.

- Quadratic. A related QARCH model of Sentana (1991) has

$$\sigma_t^2 = \alpha_0 + \alpha_1 y_{t-1}^2 + \alpha_1^* y_{t-1}. \qquad (1.11)$$

Clearly there are constraints on the parameters to ensure $\sigma_t^2 \geq 0$. Again (1.11) is used to capture asymmetry.

- Threshold. Various TARCH models have been proposed which have different parameters for $y_{t-1} > 0$ and $y_{t-1} \leq 0$. Zakoian (1990) works with the absolute residuals, while in an influential paper Glosten, Jagannathan and Runkle (1993) work with a model of the type

$$\sigma_t^2 = \alpha_0 + \alpha_1^+ I(y_t > 0) y_{t-1}^2 + \alpha_1^- I(y_{t-1} \leq 0) y_{t-1}^2.$$

  This is used by Engle and Lee (1992), who allow asymmetry to enter the transitory component of volatility, but not the permanent part.

- ARCH in mean. A theoretically important characteristic of excess returns is the relationship between expected returns of a risky asset and the level of volatility. Engle, Lilien and Robins (1987) proposed the ARCH-M model

$$y_t = g(\sigma_t^2, \theta) + \varepsilon_t \sigma_t, \quad \sigma_t^2 = \alpha_0 + \alpha_1 \{y_{t-1} - g(\sigma_{t-1}^2, \theta)\}^2.$$

  A commonly used parametrization is the linear one: $g(\sigma_t^2, \theta) = \mu_0 + \mu_1 \sigma_t^2$. Its statistical properties are studied by Hong (1991). Likelihood inference is again straightforward.

1.2.6 Simple empirical illustrations

The simple ARCH-based models are quite easy to fit to data. To illustrate this we will briefly analyze the four series introduced in section 1.1 using GARCH and EGARCH models. Throughout we will work with the compounded return on the series $y_t = 100 \log x_t / x_{t-1}$. The models will be based on Gaussian and Student t distributions where the degrees of freedom are estimated by ML techniques. When the t distribution is used the innovations from the series will be mapped into normality by using the inverse Student t distribution function followed by computing the corresponding normal deviates. These will be used as inputs into the Box–Ljung statistics (Harvey, 1993b, p. 45) for the squares using 30 lags and the standardized fourth moment, or kurtosis, statistic. The first of these statistics should be centred around 30, the second around 3.

Table 1.3 gives the results for GARCH models. To benchmark each of the results I have presented two non-ARCH models: an NID model, whose diagnostics indicate failure because of large degrees of serial correlation in the squares and fat-tails; and an independently distributed (ID) Student t model, which eliminates most of the fat tails problems, but does not deal with the correlation in the squares.

The GARCH models do improve upon these benchmarks. They have two broad effects. First, they successfully deal with the serial correlation

Table 1.3 *Each column represents the empirical fit of a specific GARCH model to the denoted series. When the parameter estimates $\hat{\alpha}_1$ and $\hat{\beta}_1$ are missing, this means they are constrained to being zero. When $\widehat{v}$, the degrees of freedom parameter, is missing, it is set to ∞ — giving a normal distribution. BL denotes the Box–Ljung statistic with 30 lags. K denotes the standardized fourth moment of the transformed innovations. K should be around 3*

		Nikkei 500 index		
$\hat{\alpha}_1$			0.198	0.161
$\hat{\beta}_1$			0.834	0.851
$\widehat{v}$		3	4	
$\log L$	−3577	−3035	−2836	−3012
BL	828	828	51.5	40.6
K	36.2	3.28	3.14	13.0
		FTSE 100 index		
$\hat{\alpha}_1$			0.116	0.100
$\hat{\beta}_1$			0.879	0.820
$\widehat{v}$		5	9	
$\log L$	−2933	−2681	−2595	−2697
BL	1375	1375	6.72	9.01
K	28.1	3.20	3.57	21.1
		German Deutsche Mark		
$\hat{\alpha}_1$			0.135	0.087
$\hat{\beta}_1$			0.902	0.896
$\widehat{v}$		3	4	
$\log L$	−1351	−1121	−945.3	−1105
BL	638	638	15.8	20.4
K	10.1	2.90	3.20	9.00
		Japanese yen		
$\hat{\alpha}_1$			0.086	0.045
$\hat{\beta}_1$			0.939	0.945
$\widehat{v}$		4	6	
$\log L$	−2084	−1983	−1879	−1937
BL	563	563	30.0	31.2
K	5.29	2.91	3.30	4.77

in the squares. Second, they reduce the fitted value of K in the normal-based model and increase the value of v in the Student t case. Both of these facts indicate that the GARCH model has explained a part of the fat tails in the distribution by a changing variance. However, I am impressed not by this result, but rather by the transpose of it. In the Nikkei 500 case, v still has to be under 4 for this model successfully to match up with the data. The other cases are nearly as extreme. Consequently, in terms of likelihood reduction, the use of the fat-tailed distribution is as important as the use of GARCH processes in modelling the data. I think this is disappointing, suggesting that GARCH models cannot deal with the extremely large movements in financial markets, even though they are good models of changing variance.

Table 1.4 *Each column represents the empirical fit of a specific EGARCH model to the denoted series. When the parameter estimate $\widehat{\omega}$ is missing, this means it is constrained to being zero. When $\widehat{v}$, the degrees of freedom parameter, is missing, it is set to ∞ — giving a normal distribution. BL denotes the Box–Ljung statistic with 30 lags. K denotes the standardized fourth moment of the transformed innovations. K should be around 3, even in the t-distribution case*

	Nikkei 500			FTSE 100		
γ_1	0.988	0.985	0.970	0.911	0.907	0.960
θ_1	−0.574	−0.478	−0.292	0.368	0.495	0.254
ω		−0.161	−0.158		−0.059	−0.031
λ	0.482	0.361	0.217	0.211	0.170	0.168
v			5			9
$\log L$	−3017	−2978	−2795	−2694	−2681	−2593
BL	130	46.2	38.6	19.2	21.3	12.9
K	10.9	18.7	3.67	17.8	13.4	3.44
	DM			**Yen**		
γ_1	0.955	0.969	0.983	0.985	0.981	0.986
θ_1	0.433	−0.139	−0.146	0.231	0.239	−0.170
ω		−0.079	−0.051		−0.027	−0.049
λ	0.154	0.176	0.232	0.089	0.080	0.141
v			4			5
$\log L$	−1119	−1104	−942	−1945	−1940	−1879
BL	19.1	22.1	15.3	33.6	32.2	30.0
K	9.75	9.27	3.61	4.78	4.76	3.20

The results of the GARCH models can be contrasted to the fit of the EGARCH models for these data sets. The results given in Table 1.4, where θ_1 denotes a moving average parameter added to equation (1.9), suggest that the use of the signs of the observations can very significantly improve the fit of the models. In this empirical work this seems to hold for both currencies and equities, although the effect is stronger in the latter. Although this is a standard result for equities (Nelson, 1991;

Brock, Lakonishok and LeBaron, 1992; Poon and Taylor, 1992), it is non-standard for currencies where the asymmetry effects are usually not significant.

1.3 Stochastic volatility

The basic alternative to ARCH-type modelling is to allow σ_t^2 to depend not on past observations, but on some unobserved components or latent structure. The most popular of these parameter-driven stochastic volatility models, from Taylor (1986), puts

$$y_t = \varepsilon_t \exp(h_t/2), \quad h_{t+1} = \gamma_0 + \gamma_1 h_t + \eta_t,$$

although the alternative parametrization $y_t = \varepsilon_t \beta \exp(h_t/2)$ and $h_{t+1} = \gamma_1 h_t + \eta_t$ also has attractions. One interpretation for the latent h_t is to represent the random and uneven flow of new information, which is very difficult to model directly, into financial markets; this follows the work of Clark (1973) and Tauchen and Pitts (1983).

For the moment ε_t and η_t will be assumed independent of one another, Gaussian white noise. Their variances will be 1 and σ_η^2, respectively. Due to the Gaussianity of η_t, this model is called a **log-normal SV model**. Its major properties are discussed in Taylor (1986; 1994). Broadly speaking these properties are easy to derive, but estimation is substantially harder than for the corresponding ARCH models.

1.3.1 Basic properties

As η_t is Gaussian, h_t is a standard Gaussian autoregression. It will be (strictly and covariance) stationary if $| \gamma_1 |< 1$ with:

$$\mu_h = E(h_t) = \frac{\gamma_0}{1-\gamma_1}, \quad \sigma_h^2 = \text{var}(h_t) = \frac{\sigma_\eta^2}{1-\gamma_1^2}.$$

As ε_t is always stationary, y_t will be stationary if and only if h_t is stationary, for y_t is the product of two stationary processes. Using the properties of the log-normal distribution, if r is even, all the moments exist if h_t is stationary and are given by:

$$E(y_t^r) = E(\varepsilon_t^r) E\left\{\exp\left(\frac{r}{2}h_t\right)\right\} \tag{1.12}$$

$$= r! \exp\left(\frac{r}{2}\mu_h + r^2\sigma_h^2/8\right) / \left(2^{r/2}(r/2)!\right). \tag{1.13}$$

All the odd moments are zero. Of some interest is the kurtosis: $E(y_t^4)/(\sigma_{y^2}^2)^2 = 3\exp(\sigma_h^2) \geq 3$. This shows that the SV model has fatter tails than the corresponding normal distribution.

The dynamic properties of y_t are easy to find. First, as ε_t is iid, y_t is an MD and is WN if $|\gamma_1| < 1$. The squares have the product moments $E(y_t^2 y_{t-s}^2) = E\{\exp(h_t + h_{t-s})\}$. As h_t is a Gaussian AR(1),

$$\begin{aligned}\operatorname{cov}(y_t^2, y_{t-s}^2) &= \exp\{2\mu_h + \sigma_h^2(1+\gamma_1^s)\} - \{E(y_t^2)\}^2 \\ &= \exp(2\mu_h + \sigma_h^2)\{\exp(\sigma_h^2\gamma_1^s) - 1\}\end{aligned}$$

and so (Taylor, 1986, pp. 74–75)

$$\rho_{y_t^2}(s) = \operatorname{cov}(y_t^2 y_{t-s}^2)/\operatorname{var}(y_t^2) = \frac{\exp(\sigma_h^2\gamma_1^s) - 1}{3\exp(\sigma_h^2) - 1} \simeq \frac{\exp(\sigma_h^2) - 1}{3\exp(\sigma_h^2) - 1}\gamma_1^s. \tag{1.14}$$

Notice that if $\gamma_1 < 0$, $\rho_{y_t^2}(s)$ can be negative, unlike the ARCH models. This is the autocorrelation function of an ARMA(1,1) process. Thus the SV model behaves in a manner similar to the GARCH(1,1) model.

The dynamic properties of the SV model can also be revealed by using logs. Clearly

$$\log y_t^2 = h_t + \log \varepsilon_t^2, \quad h_{t+1} = \gamma_0 + \gamma_1 h_t + \eta_t, \tag{1.15}$$

a linear process, which adds the iid $\log \varepsilon_t^2$ to the AR(1) h_t. As a result $\log y_t^2 \sim$ ARMA(1,1). If ε_t is normal then $\log \varepsilon_t^2$ has a mean of -1.27 and variance 4.93. It has a very long left-hand tail, caused by taking the logs of very small numbers (see Davidian and Carroll, 1987, p. 1088). The autocorrelation function for $\log y_t^2$ is

$$\rho_{\log y_t^2}(s) = \frac{\gamma_1^s}{(1 + 4.93/\sigma_h^2)}. \tag{1.16}$$

1.3.2 Estimation

The main difficulty of using SV models is that, unlike with ARCH models, it is not immediately clear how to evaluate the likelihood: the distribution of $y_t \mid Y_{t-1}$ is specified implicitly rather than explicitly. Like most non-Gaussian parameter-driven models, there are many different ways to perform estimation. Some involve estimating or approximating the likelihood, others use method-of-moments procedures.

Generalized method-of-moments (GMM)

In econometrics method-of-moments procedures are very popular. There seem to be two main explanations for this: economic theory often specifies that specific variables are uncorrelated and some econometricans are sometimes reluctant to make distributional assumptions. In the SV case neither of these explanations seems very persuasive, for the SV process is a fully specified parametric model.

In the SV case there are many possible moments to use in estimating the parameters of the model. This is because y_t^2 behaves like an ARMA(1,1) model and moving average models do not allow sufficient statistics which are of a smaller dimension than T. This suggests that the use of a finite number of moment restrictions is likely to lose information. Examples include those based on $y_t^2, y_t^4, \{y_t^2 y_{t-s}^2, s = 1, \ldots, S\}$, although there are many other possibilities. As a result, we may well want to use more moments than there are parameters to estimate, implying that they will have to be pooled. A reasonably sensible way of doing this is via generalized method of moments (GMM).

We will write, for example:

$$g_T = \begin{bmatrix} \{\frac{1}{T}\sum y_t^2\} - E(y_t^2) \\ \{\frac{1}{T}\sum y_t^4\} - E(y_t^4) \\ \\ \vdots \\ \\ \{\frac{1}{T}\sum y_t^2 y_{t-1}^2\} - E(y_t^2 y_{t-1}^2) \\ \{\frac{1}{T}\sum y_t^2 y_{t-S}^2\} - E(y_t^2 y_{t-S}^2) \end{bmatrix}$$

as the moment constraints. By varying θ, the $(S + 2) \times 1$ vector g_T will be made small. The GMM approach of Hansen (1982) suggests measuring smallness by using the quadratic form $q = g_T' W_T g_T$.The weighting matrix W_T should reflect the relative importance given to matching each of the moments. A good discussion of GMM properties is given in Hamilton (1994, Chapter 14). Earlier versions of GMM include Cramér (1946, p. 425) and Rothenberg (1973).

Applications of this method to SV include Chesney and Scott (1989), Duffie and Singleton (1993), Melino and Turnbull (1990), Jacquier, Polson and Rossi (1994), Andersen (1995) and Andersen and Sorensen (1995). These papers use a variety of moments and weighting matrices W_T. In a rather different vein, Gallant, Hsieh and Tauchen (1994) use a semi-nonparametric ARCH model to provide scores to allow the fitting by 'efficient method of moments'.

It seems to me that there are a number of obvious drawbacks to the GMM approach to the estimation of the SV model:

- GMM can only be used if h_t is stationary. When γ_1 is close to unity, as we will find for many high-frequency financial data sets, we can expect GMM to work poorly.

- Parameter estimates are not invariant. By this I mean that if the model is reparametrized as $\tau = f(\theta)$, then $\widehat{\tau} \neq f(\widehat{\theta})$. This seems important as the direct parameters in the model, γ_0, γ_1 and σ_η^2, are not fundamentally more interesting than other possible parametrizations.
- The squares y_t^2 behave like an ARMA(1,1) model. Equation (1.14) indicates that if σ_η^2 is small (as we will find in practice), $\rho_{y_t^2}(s)$ will be small but positive for many s. Even if γ_1 is close to unity, this will hold. This implies that for many series, S will have to be very high to capture the low correlation/persistence in the volatility process.
- GMM does not deliver an estimate (filtered or smoothed) of h_t. Consequently a second form of estimation will be required to carry out that task.
- Conventional tests of the time series model, based on one-step-ahead prediction densities, are not available after the fitting of the model.

Quasi-likelihood estimation

A rather simpler approach can be based on (1.15). As $\log \varepsilon_t^2 \sim$ i.i.d., $\log y_t^2$ can be written as a non-Gaussian but linear state space. Consequently the Kalman filter, given in the Appendix, can be used to provide the best linear unbiased estimator of h_t given Y_{t-1}^{l2}, where $Y_s^{l2} = (\log y_1^2, \ldots, \log y_s^2)'$. Further, the smoother gives the best linear estimator given Y_T^{l2}.

This way of estimating h_t is used by Melino and Turnbull (1990), after estimating θ by GMM. The parameters θ can also be estimated, following the suggestion of Harvey, Ruiz and Shephard (1994), using the quasi-likelihood (ignoring constants)

$$lq(\theta; y) = -\frac{1}{2}\sum_{t=1}^{T} \log F_t - \frac{1}{2}\sum_{t=1}^{T} v_t^2/F_t, \tag{1.17}$$

where v_t is the one-step-ahead prediction error and F_t is the corresponding mean squared error from the Kalman filter. If (1.15) had been a Gaussian state space then (1.17) would be the exact likelihood. As this is not true, (1.17) is called a **quasi-likelihood** and can be used to provide a consistent estimator $\widehat{\theta}$ and asymptotically normal inference. The asymptotic distribution of $\widehat{\theta}$ is discussed in Harvey, Ruiz and Shephard (1994), who use the results of Dunsmuir (1979), and relies on the usual sandwich estimator of quasi-likelihood methods.

To gain some impression of the precision of this method we report, in the first row of Figure 1.2, 500 estimates resulting from the application

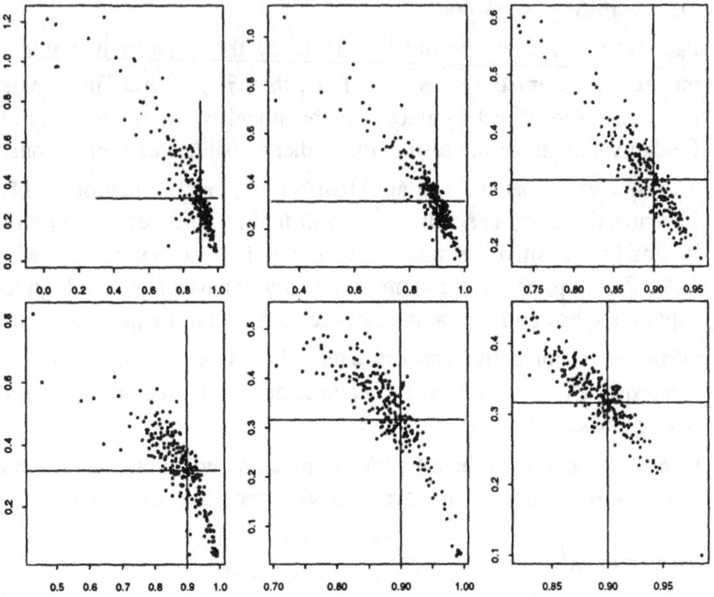

Figure 1.2 *QML and Bayes estimate of SV model. QML on top, Bayes in bottom row. Uses* $T = 500$, $T = 1000$, $T = 2000$. *T goes from left to right. The Y-axis is* $\hat{\sigma}_\eta$ *and on X-axis* $\hat{\gamma}_1$. *The crossing lines drawn on the graphs indicate the true parameter values.*

of the quasi-maximum likelihood (QML) method to simulations from the SV model with $\sigma_\eta^2 = 0.1$, $\gamma_0 = 0.0$, $\gamma_1 = 0.9$ for $T = 500, 1000$ and 2000, focusing on σ_η and γ_1. Notice the strong negative correlation between the two estimators. Later we will compare the properties of this estimator with two other likelihood suggestions. Recently, following a suggestion of Fuller (1996, Example 9.3.2), Breidt and Carriquiry (1995) have investigated modifying the $\log y_t^2$ transformation, to reduce the sensitivity of the estimation procedure to small values of y_t. Their work improves the small-sample performance of the QML estimator.

The mode

A satisfactory way of representing our knowledge of h_t is via its posterior distributions $f(h_t \mid Y_{t-1})$ and $f(h_t \mid Y_T)$. Unfortunately it is not possible to manipulate these densities into useful forms in order straightforwardly to learn their shapes. One approach to overcoming this is to report the mode of the 'smoothing density' $f(h_1, \ldots, h_T \mid y_1, \ldots, y_T)$, suggested by Durbin (1992). The approach is based on some recent work by Whittle (1991), Fahrmeir (1992) and particularly Durbin and Koopman (1992) and Durbin (1996).

The mode of $h \mid Y_T$ is the mode of the joint density of h, Y_T. The log of this density, setting $\gamma_0 = 0$ and $h_1 = 0$ for simplicity, is:

$$l = -\frac{1}{2}\sum_{t=1}^{T}\{y_t^2 \exp(-h_t) + h_t + (h_{t+1} - \gamma_1 h_t)^2/\sigma_\eta^2\}. \qquad (1.18)$$

Then $\partial l/\partial h_t$ is nonlinear in h and so we have to resort to solving $\partial l/\partial h = 0$ iteratively. The standard way of carrying this out (ignoring the Gaussian part due to the transition equation) is to start off the kth iteration at $h_1^{(k)}, \ldots, h_T^{(k)}$ and write

$$\begin{aligned}
\frac{\partial \log f(y_t|h_t)}{\partial h_t} &= \frac{\partial \log f(y_t|h_t)}{\partial \exp(-h_t)}\frac{\partial \exp(-h_t)}{\partial h_t} \\
&= \left\{y_t^2 - \exp(h_t)\right\}\exp(-h_t)/2 \\
&\simeq \left\{y_t^2 - \exp\left(h_t^{(k)}\right)\right. \\
&- \left.\left(h_t - h_t^{(k)}\right)\exp\left(h_t^{(k)}\right)\right\}\exp\left(-h_t^{(k)}\right)/2.
\end{aligned}$$

This linearized derivative has the same form as a Gaussian measurement model, with

$$\begin{aligned}
y_t^2 - \exp\left(h_t^{(k)}\right) - h_t^{(k)}\exp\left(h_t^{(k)}\right) &= \exp\left(h_t^{(k)}\right)h_t + \varepsilon_t, \varepsilon_t \\
&\sim N\left[0, 2\exp\left(h_t^{(k)}\right)\right].
\end{aligned}$$

Hence the Kalman filter and analytic smoother (see the Appendix for details), applied to this model, solves the linearized version of the approximation to $\partial l/\partial h_t = 0$. Repeated uses of this approximation will converge to the joint mode as (1.18) is concave in h_t.

Importance sampling

A more direct way of performing inference is to compute the likelihood, integrating out the latent h_t process:

$$f(y_1, \ldots, y_T) = \int f(y_1, \ldots, y_T \mid h) f(h) dh. \qquad (1.19)$$

As this integral has no closed form it has to be computed numerically, integrating over $T \times \dim(h_t)$ dimensional space, which is a difficult task. One approach to this problem is to use Monte Carlo integration, say by drawing from the unconditional distribution of h, with the

jth replication being written as h^j, and computing the estimate $(1/M)\sum_{j=1}^{M} f(y_1,\ldots,y_T \mid h^j)$. This is likely to be a very poor estimate, even with very large M. Consequently it will be vital to use an importance sampling device (see Ripley, 1987, p. 122) to improve its accuracy. This rewrites (1.19) as

$$f(y) = \int \frac{f(y \mid h)f(h)}{g(h \mid y)} g(h \mid y)dh,$$

where it is easy to draw from some convenient $g(h \mid y)$. Replications from this density will be written as h^i, giving a new estimate, $(1/M)\sum_{i=1}^{M} f(y \mid h^i)f(h^i)/g(h^i \mid y)$. In some very impressive work, Danielsson and Richard (1993) and Danielsson (1994) have designed g functions which recursively improve (or accelerate) their performance, converging towards the optimal g_*. The details will not be dealt with here as they are quite involved even for the simplest model.

1.3.3 Markov chain Monte Carlo

Although importance sampling shows promise, it is likely to become less useful as T becomes large or $\dim(h_t)$ increases beyond 1. For such difficult problems a natural approach is one based on Markov chain Monte Carlo (MCMC). MCMC will be used to produce draws from $f(h \mid y)$ and sometimes, if a Bayesian viewpoint is taken, the posterior on the parameters $\theta \mid y$. For the moment we will focus on the first of these two targets.

Early work on using MCMC for SV models focused on **single-move** algorithms, drawing h_t individually, ideally from its conditional distribution $h_t \mid h_{\setminus t}, y$, where the notation $h_{\setminus t}$ means all elements of h except h_t. However, a difficulty with this is that although

$$\begin{aligned} f(h_t \mid h_{\setminus t}, y) &= f(h_t \mid h_{t-1}, h_{t+1}, y_t) \\ &\propto f(y_t \mid h_t)f(h_{t+1} \mid h_t)f(h_t \mid h_{t-1}), \quad (1.20) \end{aligned}$$

has an apparently simple form, the constant of proportionality is unknown. As a result it seemed difficult to sample directly from (1.20). Shephard (1993) used a random walk Metropolis algorithm to overcome this problem. A much better approach is suggested in Jacquier, Polson and Rossi (1994), building on the work of Carlin, Polson and Stoffer (1992).

Rejection Metropolis

Jacquier, Polson and Rossi (1994) suggest using a Metropolis algorithm built around an accept/reject kernel, which uses an approximation to

(1.20) which is easy to simulate from, written as $g(h_t \mid h_{\backslash t}, y)$. Then if there exists a c such that

$$f(h_t \mid h_{\backslash t}, y) \leq cg(h_t \mid h_{\backslash t}, y), \quad \forall h_t, \tag{1.21}$$

we could sample from f by drawing from g and accepting this with probability $f(h_t \mid h_{\backslash t}, y) / \{g(h_t \mid h_{\backslash t}, y)c\}$.

Jacquier, Polson and Rossi (1994) argued that it was difficult to find a valid g satisfying (1.21) for all h_t, but that this is not so important as we can use a g inside a Metropolis algorithm which overcomes this approximation. Their proposal takes on the form of moving from h_t to h_t^* with probability

$$\min\left[1, \frac{f(h_t^* \mid h_{\backslash t}, y)}{f(h_t \mid h_{\backslash t}, y)} \frac{\min\{f(h_t \mid h_{\backslash t}, y), cg(h_t \mid h_{\backslash t}, y)\}}{\min\{f(h_t^* \mid h_{\backslash t}, y), cg(h_t^* \mid h_{\backslash t}, y)\}}\right].$$

Note that our lack of knowledge of the constant of proportionality in (1.20) is now irrelevant as it cancels in this expression. The choice of g governs how successful this algorithm will be. If it is close to f, c can be close to 1 and the algorithm will almost always accept the moves. Jacquier, Polson and Rossi (1994) suggest using an inverse gamma distribution to approximate the distribution of $\exp(h_t) \mid h_{t-1}, h_{t+1}, y_t$ and a variety of *ad hoc* rules for selecting c.

In my discussion of single-move MCMC I am going to avoid using the Jacquier, Polson and Rossi (1994) method as I think there are now simpler ways of proceeding. One approach is to devise an accept/reject algorithm based around the prior. We write $h_t | h_{t-1}, h_{t+1} \sim N(h_t^*, \sigma_t^2)$, then $\log f(h_t | y_t, h_{t-1}, h_{t+1}) = \text{const} + \log f^*$ where

$$\begin{aligned} \log f^* &= -\frac{1}{2}h_t - \frac{1}{2\sigma_t^2}(h_t - h_t^*)^2 - \frac{1}{2}\left\{y_t^2 \exp(-h_t)\right\} & (1.22) \\ &\leq -\frac{1}{2}h_t - \frac{1}{2\sigma_t^2}(h_t - h_t^*)^2 & (1.23) \\ &\quad - \left(\frac{y_t^2}{2}\right)\{\exp(-h_t^*)(1 + h_t^*) - h_t \exp(-h_t^*)\} & (1.24) \\ &= \log g^* & (1.25) \end{aligned}$$

is a bounding function. Hence it is a trivial matter to draw from f using an accept/reject algorithm. The proposal, drawn from the normalized version of g^*, a normal distribution, has mean and variance

$$\mu_t = h_t^* + \frac{\sigma_t^2}{2}\left[y_t^2 \exp(-h_t^*) - 1\right] \quad \text{and} \quad \sigma_t^2 = \sigma_\eta^2 / \left(1 + \gamma_1^2\right).$$

Hence we can sample from $h_t \mid h_{t-1}, h_{t+1}, y_t$ by proposing $h_t \sim N(\mu_t^*, \sigma_t^2)$ and accepting with probability f^*/g^*. This idea, suggested and extended in Pitt and Shephard (1995), gave a 99.5% acceptance rate in a Monte Carlo study using the true values $\gamma_1 = 0.9, \sigma_\eta^2 = 0.05, \gamma_0 = 0$, while it executed about 10 times faster than the code supplied by Jacquier, Polson and Rossi. It is possible also to include the second-order term in the Taylor expansion of $\exp(-h_t)$. This naturally leads to a Metròpolis sampler (as it no longer bounds) with a very small rejection probability. Finally, the second-order expansion can be extended to allow a multimove algorithm.

An alternative to this algorithm is to note that $f^*(h_t \mid h_{t-1}, h_{t+1}, y_t)$ is log-concave, which means that the adaptive routines of Gilks and Wild (1992) and Wild and Gilks (1993) can be used.

Whichever of these two algorithms is used, we can now use the Gibbs sampler:

1. Initialize h.
2. Draw h_t^* from $f(h_t \mid h_{t-1}^*, h_{t+1}, y_t), \quad t = 1, \ldots, T$.
3. Write $h = h^*$; go to 2.

This sampler will converge to drawings from $h \mid y$ so long as $\sigma_\eta^2 > 0$. However, the speed of convergence may be slow, in the sense of taking a large number of loops. To illustrate these features, Figure 1.3 reports two sets of experiments, each computed using two sets of parameter values: $\gamma_1 = 0.9$, $\sigma_\eta^2 = 0.1$ and $\gamma_1 = 0.99$, $\sigma_\eta^2 = 0.01$. The first experiment reports the runs of 500 independent Gibbs samplers, each initialized at $h_t = 0$ for all t, as they iterated. The lines are the average of the samplers after given numbers of iterations. This experiment is designed to show how long the initial conditions last in the sampler and so reflect the memory or correlation in the sampler. The second experiment runs a single Gibbs sampler for 100 000 iterations, discarding the first 10 000 results, for the above two sets of parameter values. The resulting 90 000 drawings from $h_{50}|Y_{100}$ were inputted into a correlogram and are reported in Figure 1.3. The idea is to represent the correlation in the sampler once it has reached equilibrium.

The results of Figure 1.3 are very revealing for they show that as γ_1 increases (and similarly as $\sigma_\eta^2 \to 0$), so the sampler slows up, reflecting the increased correlation among the $h \mid y$. This unfortunate characteristic of the single-move Gibbs sampler is common to all parameter-driven models (see Carter and Kohn, 1994; Shephard, 1994b). If a component, such as h_t, changes slowly and persistently, the single-move sampler will be slow. In the limit, when $h_t = h_{t-1}$, the sampler will not converge at all. Given that volatility tends to move slowly, this suggests that this algorithm may be unreliable for real finance problems.

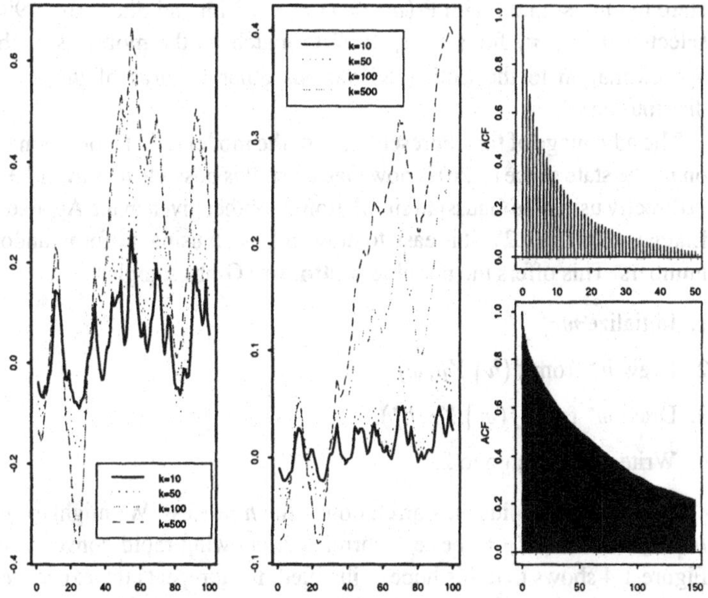

Figure 1.3 *Signal extraction of SV model using single-move Gibbs. Indicates rate of convergence. Left picture is* $\gamma_1 = 0.9$ *case. Middle gives* $\gamma_1 = 0.99$. *Correlograms: Top has* $\gamma_1 = 0.9$, *bottom has* $\gamma_1 = 0.99$. *In the first two graphs the X-axis is t, and the Y-axis is an estimate of* $E(h_t|Y_T)$.

Multimove samplers

A common solution in the MCMC literature to the problem of elements of $h \mid y$ being highly correlated is to avoid sampling single elements of $h_t \mid h_{\setminus t}, y$, by working with blocks (see Smith and Roberts, 1993, p. 8; Liu, Wong and Kong, 1994). In the context of time series models, early work on designing methods to sample blocks includes that by Carter and Kohn (1994) and Fruhwirth-Schnatter (1994) which has now been refined by de Jong and Shephard (1995). This work can be used to analyse the SV model by using the linear state-space representation:

$$\log y_t^2 = h_t + \log \varepsilon_t^2, \quad \varepsilon_t \sim NID(0,1). \tag{1.26}$$

The idea, suggested in Shephard (1994b) and later used by Carter and Kohn (1994) and Mahieu and Schotman (1994), is to approximate the $\log \varepsilon_t^2$ distribution by a mixture of normals so that:

$$\log \varepsilon_t^2 \mid w_t = j \sim N(\mu_j, \sigma_j^2), \quad j = 1, \ldots, J. \tag{1.27}$$

Here the $w_t \sim$ i.i.d., with $\Pr(w_t = j) = \pi_j$. Kim and Shephard (1994) selected μ_j, σ_j^2, π_j for $j = 1, \ldots, 7$ to match up the moments of this approximation to the truth (and various other features of the $\log \chi_1^2$ distribution).

The advantage of this representation of the model is that, conditionally on w, the state space (1.26) is now Gaussian. It is possible to draw $h \mid Y_T$, w directly using the Gaussian simulation smoother given in the Appendix. Likewise, using (1.27) it is easy to draw $w \mid Y_T, h$ using uniform random numbers. This offers the possible **multimove** Gibbs sampler:

1. Initialize w.
2. Draw h^* from $f(h \mid Y_T, w)$.
3. Draw w^* from $f(w \mid Y_T, h^*)$.
4. Write, $w = w^*$; go to 2.

This sampler avoids the correlation in the h process. We might expect $h \mid Y_T$ and $w \mid Y_T$ to be less correlated, allowing rapid convergence. Figure 1.4 shows that this hope is justified, for it repeats the experiment of Figure 1.3 but now uses a multimove sampler. Convergence to the equilibrium distribution appears more rapid, while there is substantially less correlation in the sampler once equilibrium is obtained.

Although the multimoving can be carried out by transforming the model and using a mixture representation, it could be argued that this is only an approximation. It is a challenging problem to come up with multimove algorithms without transforming the model since a fast and, more importantly, reliable sampler will improve the usefulness of these MCMC techniques.

Bayesian estimation

The ability to sample from $h \mid Y_T$ means parameter estimation is reasonably straightforward. The simplest approach to state is the Bayesian one: it assumes a known prior, $f(\theta)$, for $\theta = (\sigma_\eta^2, \gamma_1)$. Then the multimove sample, for example, becomes, when we write $g(h \mid Y_T, \theta)$ to denote a MCMC update using either a Gibbs sampler or multimove sampler:

1. Initialize θ.
2. Draw h^* from $g(h \mid Y_T, \theta)$.
3. Draw θ^* from $f(\theta \mid h^*)$,
4. Write $\theta = \theta^*$; go to 2.

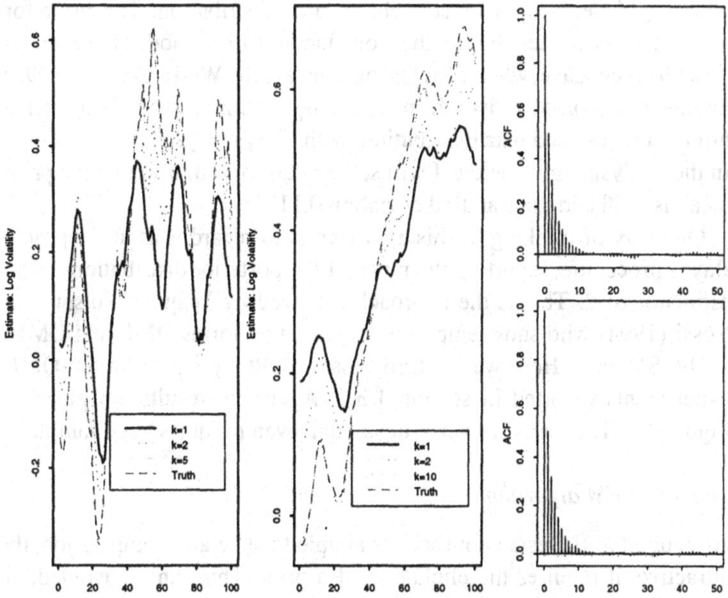

Figure 1.4 *Signal extraction of SV model using multimove Gibbs. Indicates rate of convergence. Left picture is* $\gamma_1 = 0.9$ *case. Middle gives* $\gamma_1 = 0.99$. *The right hand picture indicates serial correlation in the sampler. The correlogram on the top corresponds to the low-persistence case, the one on the bottom represents the high-persistence case. In the first two graphs the X-axis is t, and the Y-axis is an estimate of* $Eh_t|Y_T$. *Truth denotes the actual value of* $Eh_t|Y_T$.

As the likelihood $f(h^* \mid \theta)$ is Gaussian it is tempting to use the standard normal-inverse gamma conjugate prior for θ, as do Jacquier, Polson and Rossi (1994). However, I think there are advantages in enforcing the stationarity conditions on γ_1, and hence on the h process, which a Gaussian prior on γ_1 will not achieve. It can be carried out by dividing step 3 into two parts:

3a. Draw σ_η^{2*} from $f(\sigma_\eta^2 \mid h^*, \gamma_1)$.
3b. Draw γ_1^* from $f(\gamma_1 \mid h^*, \sigma_\eta^{2*})$.

The likelihood $f(h \mid \theta)$ suggests a simple non-informative conjugate prior for $\sigma_\eta^2 \mid h, \gamma_1$ yielding the posterior

$$\chi_T^{-2}\left\{\sum_{t=2}^{T}(h_t - \gamma_1 h_{t-1})^2 + h_1^2(1-\gamma_1^2)\right\},$$

where χ^{-2} denotes an inverse chi-squared distribution. The prior for $\gamma_1 \mid h, \sigma_\eta^2$ is harder due to the non-standard likelihood. However, as $f(h \mid \theta)$ is concave, when $f(\gamma_1)$ is log-concave the Wild and Gilks (1993) method can be used to draw from $\gamma_1 \mid h, \sigma_\eta^2$. A simple example of such a prior is the rescaled beta distribution, with $E(\gamma_1) = \{2\alpha/(\alpha+\beta)\} - 1$. In the analysis I give below, I will set $\alpha = 20$, $\beta = 3/2$ so that the prior mean is 0.86 and the standard deviation 0.11.

One way of thinking of this approach is to regard it as an empirical Bayes procedure, reporting the mean of the posterior distributions as an estimator of θ. This is the approach followed by Jacquier, Polson and Rossi (1994) who show empirical Bayes outperforms QML and GMM in the SV case. Here we confirm those results by repeating the QML experiments reported in section 1.3.2. Again the results are given in Figure 1.2. The gains are very substantial, even for quite large samples.

Simulated EM algorithm

Although the Bayesian approach is simple to state and computationally attractive, it requires the elicitation of a prior. This can be avoided, at some cost, by using MCMC techniques inside a simulated EM algorithm. This was suggested for the SV model by Kim and Shephard (1994). An excellent introduction to the statistical background for this procedure is given in Qian and Titterington (1991); see also the more recent work by Chan and Ledolter (1995). Earlier work on this subject includes Bresnahan (1981), Wei and Tanner (1990) and Ruud (1991).

The EM algorithm works with the mixture-of-normals representation given above, where w is the mixture number. Then

$$\log f(Y_T;\theta) = \log f(Y_T \mid w;\theta) + \log \Pr(w) - \log \Pr(w \mid Y_T;\theta).$$

As $Y_T \mid w$ is a Gaussian state space, its log-density can be evaluated using the Kalman filter

$$\log f(Y_T \mid w;\theta) = \text{const} - \frac{1}{2}\sum_{t=1}^{T} \log F_t - \frac{1}{2}\sum_{t=1}^{T} v_t^2/F_t.$$

As $f(w)$ is parameter-free, the next step of the EM algorithm is found as:

$$\theta^{(i+1)} = \arg\max_{\theta} \sum_{w} \log f(Y_T \mid w;\theta)\Pr(w \mid Y_T;\theta^{(i)}).$$

As $\Pr(w \mid Y_T;\theta^{(i)})$ is unknown, it is not possible to solve this maximization directly. It is replaced by an estimate, using simulations from $\Pr(w \mid Y_T;\theta^{(i)})$ drawn using MCMC techniques. Consequently,

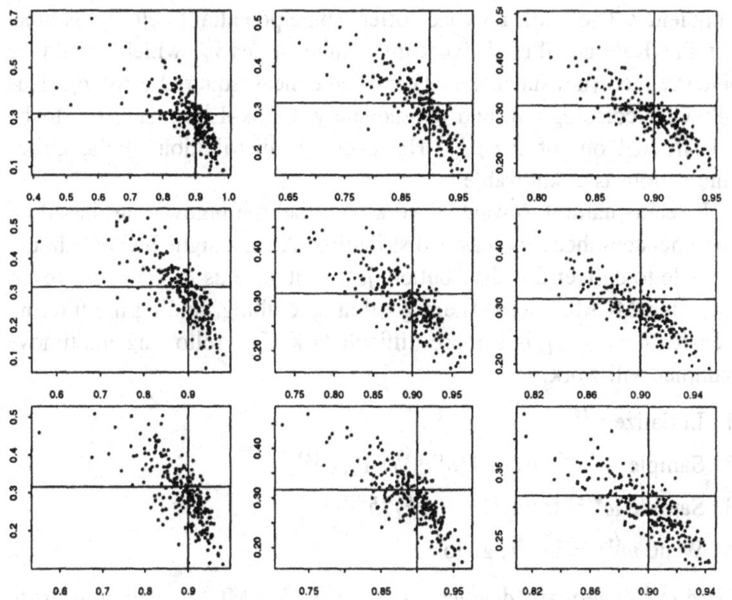

Figure 1.5 *Simulating EM estimate of SV model. Uses* $T = 500$, $T = 1000$, $T = 2000$ *and* $M = 1$, $M = 3$, $M = 10$. *T goes from right to left.* $M = 1$ *in the top row. The Y-axis is* $\hat{\sigma}_\eta$ *and the X-axis* $\hat{\gamma}_1$. *The crossing lines drawn on the graphs indicate the true parameter values.*

the function which is numerically maximized is $(1/M)\sum_{i=1}^{M} \log f(Y_T \mid w^i; \theta)$. As $M \to \infty$ so this algorithm converges to the EM algorithm. It may be possible to construct an asymptotic theory for the resulting iterated estimator even if M is finite, using the simulated scores argument of Hajivassiliou and McFadden (1990).

There is some hope that this EM algorithm will converge very quickly to the ML estimator since the 'missing' data w is not very informative about θ. The results, using $M = 1, 3, 10$ and employing 10 steps of the EM algorithm, are reported in Figure 1.5. Kim and Shephard (1994) have found little gain in taking M much bigger than 10, although for more complicated models the situation may be different.

Diagnostic checking

Although there is now a vast literature on fitting SV models, there is barely a word on checking them formally; a notable exception to this is the paper by Gallant, Hsieh and Tauchen (1994). This is an important deficiency. The QML approach offers some potential to close this hole, for the Kalman filter delivers quasi-innovations $\widetilde{v}_t$, which should be uncorrelated (not independent) and have mean squared error F_t. This allows a correlogram, and consequently a Box–Ljung statistic, to be constructed out of $\widetilde{v}_t/\sqrt{F_t}$. However, the distribution of the quasi-innovations is unknowable.

It seems natural to want to work with the true innovations, based on the one-step-ahead forecast's distribution. At first sight MCMC should be able to deliver this distribution, just as it gave us $h_t \mid Y_T$. However, although MCMC methods are good at smoothing, finding the filtering density $h_t \mid Y_{t-1}$ is a more difficult task. The following multimove sampler will work:

1. Initialize $w^{(t)}$.
2. Sample $h^{*(t+1)}$ from $f(h^{(t+1)} \mid Y_t, w^{(t)})$
3. Sample $w^{*(t)}$ from $f(w^{(t)} \mid Y_t, h^{*(t)})$
4. Write $w^{(t)} = w^{*(t)}$; go to 2.

Here $r^{(t)}$ generically denotes $(r_1, \ldots, r_t)$. This MCMC will allow us to sample from $h_{t+1} \mid Y_t$ and so estimate:

$$\Pr(y_{t+1} \leq x \mid Y_t) = F_{y_{t+1}|Y_t}(x) \doteq \frac{1}{M}\sum_{i=1}^{M} \Pr(y_{t+1} \leq x \mid h_{t+1}^{(i)}).$$

These distribution functions or probabilities are vital for they provide the natural analogue of the Gaussian innovations from a time series model. The first reference I know to them is Smith (1985) who noted that it is possible to map them into any convenient distribution to allow easy diagnostic checking. Examples of their use will be given later in section 1.3.5.

A significant difficulty with the MCMC approach is that if T is large it will be computationally expensive. The diagnostic simulation is $O(T^2)$, which is unsatisfactory. Some work on avoiding this has been carried out by Berzuini *et al.* (1994) and Geweke (1994). More work needs to be carried out on this important topic.

1.3.4 Extensions of SV

The basic log-normal SV model can be generalized in a number of directions. A natural framework might be based on adapting the Gaussian state space so that

$$y_t = \varepsilon_t \exp\{z_t' h_t/2\}, \quad h_{t+1} = T_t h_t + \eta_t, \quad \eta_t \sim N(0, H_t).$$

A straightforward generalization might allow h_{t+1} to follow a more complicated ARMA process. Perhaps more usefully, inspiration for new components can be found in the linear models of Harrison, West and Harvey. A simple example would be

$$z_t = \begin{pmatrix} 1 \\ 1 \end{pmatrix}, \quad h_{t+1} = \begin{pmatrix} \gamma_1 & 0 \\ 0 & 1 \end{pmatrix} h_t + \eta_t,$$

where

$$\eta_t \sim N\left\{0, \begin{pmatrix} \sigma_\eta^2 & 0 \\ 0 & \sigma_p^2 \end{pmatrix}\right\}.$$

Now h_{2t+1} is a random walk, allowing the permanent level of the volatility to change slowly. This is analogous to the Engle and Lee (1992) decomposition of shocks into permanent and transitory. A model along the same lines has been suggested by Harvey and Shephard (1993a) and Carter and Kohn (1993), who allow (ignoring the cyclical AR(1) component):

$$z_t = \begin{pmatrix} 1 \\ 0 \end{pmatrix}, \quad T_t = \begin{pmatrix} 1 & 1 \\ 0 & 1 \end{pmatrix}, \quad H_t = \begin{pmatrix} 0 & 0 \\ 0 & \sigma_p^2 \end{pmatrix}.$$

This uses the Kitagawa and Gersch (1984) 'smooth trend' model in the SV context, which in turn is close to putting a cubic spline through the data. This may provide a good summary of historical levels of volatility, but it could be poor as a vehicle for forecasting as confidence intervals for forecasted h_{T+s} may grow very quickly with s. Another suggestion is to allow h_t to be a fractional process, giving the SV model long memory. This has been discussed by Harvey (1993a) and Breidt, Crato and de Lima (1993).

Asymmetric response

One motivation for the EGARCH model introduced by Nelson (1991) was to capture the non-symmetric response of the condition to shocks. A similar feature can be modelled using an SV model by allowing ε_t and η_t to be correlated. Notice ε_t is correlated with η_t, not η_{t-1}. The former model is an MD, the latter is not. If ε_t and η_t are negatively correlated, and if $\varepsilon_t > 0$, then $y_t > 0$ and h_{t+1} is likely to fall. Hence, a large y_t^2's effect on the estimated h_{t+1} will be accentuated by a negative sign on y_t, while its effect will be partially ameliorated by a positive sign.

This correlation between ε_t and η_t was suggested by Hull and White (1987) and estimated using GMM by Melino and Turnbull (1990) and Scott (1991). A simple quasi-likelihood method has been proposed recently by Harvey and Shephard (1993b). Jacquier, Polson and

Rossi (1995) have extended their single-move MCMC sampler to estimate this effect.

Finally, the Engle, Lilien and Robins (1987) ARCH-M model can be extended to the SV framework, by specifying $y_t = \mu_0 + \mu_1 \exp(h_t) + \varepsilon_t \exp(h_t/2)$. This model allows y_t to be moderately serially correlated. It is analysed in some depth by Pitt and Shephard (1995).

1.3.5 Simple empirical applications

To provide a simple illustration of the use of SV models we will repeat the analysis of ARCH-type models presented in the previous section. The SV models used will be the simple AR(1) log-normal-based process, with Gaussian measurement error. We will use a simulated EM algorithm and an empirical Bayes procedure to perform the estimation and use the diagnostic simulator to produce diagnostic checks.

Note that there is a problem with computing the ARCH likelihood. Usually the likelihood for ARCH is found by conditioning on some initial observations. In the previous section the ARCH likelihood was formed by conditioning on 20 initial observations to find σ_{20}^2, and computing a prediction decomposition using the observations from time index 21 to T. In Tables 1.5 and 1.6 we used the unconditional variance to initialize σ_0^2. This technique was used to make the computed likelihood comparable with that for the SV model. The SV model has a properly defined density $f(y_1, \ldots, y_T)$, as h_0 has a proper unconditional distribution. This accounts for the difference in the ARCH likelihoods reported in Tables 1.3, 1.5 and 1.6.

The fitted models are reported in Tables 1.5 and 1.6 and posterior distributions for the parameters are given in Figure 1.6. The approximate symmetry of the posteriors for γ_1 and β means that the empirical Bayes and simulated EM algorithms give very similar results for those parameters. The variance parameter, σ_η has a noticeable right-hand tail and so the result of the empirical Bayes solution having a higher value than the simulated EM algorithm is not surprising.

The results, given in Tables 1.5 and 1.6, suggest that the SV models are empirically more successful than the normal-based GARCH models. This should provide some assurance for option pricing theorists who price assets using this very simple SV model. However, the success of the SV model is accounted for by its better explanation of the fat-tailed behaviour of returns: SV is not a better model of volatility, it is a better model of the distribution of returns. The use of a t-distribution GARCH model overturns this SV outperformance, but not dramatically. The diagnostics of these two models seem similar. There is some evidence that the use

Table 1.5 *Empirical fits of SV models. BL denotes the Box–Ljung statistic with 30 lags. K denotes the standardized fourth moment. ARCH denotes the likelihood of the best normal ARCH models. The ARCH model is initialized using $\sigma_0^2 = \alpha_0/(1-\alpha_1-\beta_1)$. The SV model is initialized by the unconditional distribution of h_0. CI denotes a 95% Bayesian confidence interval*

SV	FTSE 100		
	SIEM	Bayes	CI
γ_1	0.945	0.944	[0.903, 0.962]
σ_η	0.212	0.215	[0.116, 0.269]
β	−0.452	−0.446	[−0.660,−0.054]
$\log L$	−2651		
BL	46.1		
K	5.05		
ARCH			
$\alpha_1+\beta_1$	0.921	0.944	
v		9	
$\log L$	−2725	−2623	
BL	9.0	22.6	
K	21.2	3.21	
SV	Yen		
	SIEM	Bayes	CI
γ_1	0.967	0.957	[0.932, 0.977]
σ_η	0.190	0.223	[0.172, 0.280]
β	−1.14	−1.16	[−0.914,−1.393]
$\log L$	−1929		
BL	36.5		
K	3.67		
ARCH			
$\alpha_1+\beta_1$	0.994	0.997	
v		5	
$\log L$	−1963	−1903	
BL	31.4	37.5	
K	4.81	4.00	

Table 1.6 *Empirical fits of SV models. BL denotes the Box–Ljung statistic with 30 lags. K denotes the standardized fourth moment. ARCH denotes the likelihood of the best normal ARCH models. The ARCH model is initialized using* $\sigma_0^2 = \alpha_0/(1-\alpha_1-\beta_1)$*. The SV model is initialized by the unconditional distribution of* h_0*. CI denotes a 95% Bayesian confidence interval*

SV	Nikkei		
	SIEM	Bayes	CI
γ_1	0.936	0.936	[0.915, 0.968]
σ_η	0.424	0.426	[0.167, 0.261]
β	−0.363	−0.359	[−0.631,−0.260]
$\log L$	−2902		
BL	88.3		
K	5.51		
ARCH			
$\alpha_1+\beta_1$	0.989	0.978	
v		4	
$\log L$	−3036	−2853	
BL	48.4	34.3	
K	13.5	3.00	
SV	DM		
	SIEM	Bayes	CI
γ_1	0.951	0.947	[0.924, 0.967]
σ_η	0.314	0.333	[0.276, 0.390]
β	−2.08	−2.09	[−2.36,−1.81]
$\log L$	−1007		
BL	28.1		
K	4.08		
ARCH			
$\alpha_1+\beta_1$	0.986	0.998	
v		4	
$\log L$	−1127	−968	
BL	19.2	31	
K	9.07	2.96	

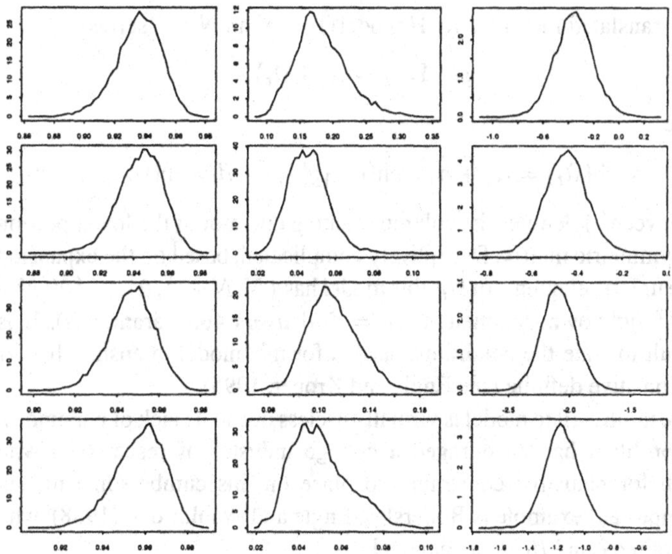

Figure 1.6 *Estimated posterior densities for* γ_1, σ_η *and* β *using the multimove Gibbs sampler. Top picture corresponds to the FTSE 100 case, then Nikkei 500, DM and yen.*

of a fat-tailed distribution on either ε_t or, the way I would prefer, on η_t would improve the fit of the SV model. Finally, fitting more complicated SV models, such as ones based on an ARMA(1,1) h_t process, could be empirically more successful for some assets.

An interesting stylized fact to emerge from this table is that the estimated γ_1 parameter is typically lower for SV models than the corresponding $\alpha_1 + \beta_1$ for GARCH. This fact has yet to be explained.

1.4 Multivariate models

Most of macro-economics and finance is about how variables interact, which, for multivariate volatility models, means it is important to capture changing cross-covariance patterns. Multivariate modelling of means is difficult and rather new: constructing multivariate models of covariance is much harder, dogged by extreme problems of lack of parsimony.

1.4.1 Multivariate ARCH

Multivariate ARCH models have existed nearly as long as the ARCH model itself. Kraft and Engle (1982) introduced the basic model which (here translated into a GARCH model) has for an $N \times 1$ series

$$y_t \mid Y_{t-1} \sim N(0, \Omega_t),$$

where

$$\text{vech}(\Omega_t) = \alpha_0 + \alpha_1 \text{vech}(y_{t-1} y_{t-1}') + \beta_1 \text{vech}(\Omega_{t-1}),$$

where vech$(\cdot)$ denotes the column stacking operator of the lower portion of a symmetric matrix. Deceptively complicated, based on the expansion of the unique elements of Ω_t, this model has $\{N(N+1)/2\} + 2\{N(N+1)/2\}^2$ unknown parameters ($N = 5$ delivers 465 parameters). It is difficult to state the conditions needed for this model to ensure that Ω_t stays positive definite (see Engle and Kroner, 1995).

The multivariate model is virtually useless due to its lack of parsimony. The problem has encouraged a cottage industry of researchers who search for plausible constraints to place on this cumbersome model. An important example is Bollerslev, Engle and Wooldridge (1988), who constrain α_1 and β_1 to be diagonal.

Constant correlation matrix

One of the more empirically successful multivariate ARCH models is the constant correlation model of Bollerslev (1990), who allows the (i, j)th element of Ω_t, Ω_{tij}, to be

$$\Omega_{tij} = \rho_{ij} h_{iit}^{1/2} h_{jjt}^{1/2}, \text{ where } h_{iit} = \alpha_{0i} + \alpha_{1i} y_{it-1}^2 + \beta_{1i} h_{iit-1}.$$

This highly constrained model implies that corr$(y_{it}, y_{jt} \mid Y_{t-1})$ is constant over time. This is often found to be empirically reasonable (see Baillie and Bollerslev, 1990), but it does lack the flexibility required to address some interesting theoretical finance issues which relate to the importance of changing correlation.

1.4.2 Multivariate asset returns

The above multivariate models are either extremely unparsimonious or quite tightly constrained. It seems useful to see if we can look to economic theory to guide us in constructing some more useful models. To start with, I will follow King, Sentana and Wadhwani (1994) and work with an $N \times 1$ series of excess returns (over a riskless interest rate),

$$y_t = \mu_t + \eta_t.$$

Here, given some common information set z_t (perhaps lagged observations or some latent process), μ_t is the expected return of the asset above a safe interest rate, and η_t is the corresponding unexpected component. The covariance of returns will be modelled using a factor structure (see Bartholomew, 1987, pp. 8–9) for η_t, with

$$\eta_t = \sum_{j=1}^{K} b_j f_{jt} + v_t = Bf_t + v_t.$$

Here $f_{1t}, \ldots, f_{Kt}$ and v_t will be assumed independent of one another. Then

$$\text{var}(\eta_t \mid z_t) = B\Lambda_t B' + \Psi_t,$$

where

$$\text{var}(f_{jt} \mid z_t) = \sigma_{jt}^2 \text{ and } \Lambda_t = \text{diag}(\sigma_{1t}^2, \ldots, \sigma_{Kt}^2).$$

The Ψ_t will be assumed to be diagonal and sometimes time-invariant. The $N \times (N-K)$ matrix of weights, B, will be called the factor loadings.

This framework can reveal that the covariance structure of the N assets influences the returns μ_t by using the arbitrage pricing theory from Ross (1976). It states that $\mu_t = B\pi_t$, where π_t is a vector whose jth element is the risk premium of a portfolio made up entirely of factor f_{jt}. Hence the risk premium of an asset is a linear combination of the risk premiums on the factors.

Unfortunately the Ross (1976) theory does not tell us how to measure risk premiums, although most finance theorists would put the risk premiums π_t as linear combinations of Λ_t. This has been justified in a formal setting by Hansen and Singleton (1983) in their work on consumption-based asset pricing theory; see also the Appendix of Engle, Ng and Rothschild (1990). In either case this delivers the model for asset returns

$$y_t = B\Lambda_t \tau + Bf_t + v_t, \tag{1.28}$$

where τ is a $K \times 1$ vector of constants. In the univariate model this delivers the ARCH-M and SV-M models outlined in the previous two sections.

From the econometrician's viewpoint (1.28) is a rather incomplete model, as z_t is unspecified. However, it can be completed by using observation-driven or parameter-driven processes, leading to factor ARCH and SV models. In this section the risk premium term $B\Lambda_t\tau$ will tend to be dropped for expositional reasons.

1.4.3 Factor ARCH models

The basis of the factor ARCH model will be

$$y_t = Bf_t + \varepsilon_t, \text{ and } f_{it} \mid Y_{t-1} \sim N(0, \sigma_{it}^2), \quad i = 1, \ldots, K,$$

where $\varepsilon_t \sim NID(0, \Psi)$. The important feature of the model is that $f_t = (f_{it}, \dots, f_{Kt})'$ is observation-driven, i.e. conditionally on Y_{t-1} its distribution is specified for this delivers the likelihood.

This model, introduced by Engle, Ng and Rothschild (1990), potentially improves the parsimony problem if a good mechanism for σ_{it}^2 can be found, for now the complexity of the model is really only of the dimension of f_t. To understand the way Engle, Ng and Rothschild (1990) suggest driving σ_{it}^2, it is useful to study briefly some of the features of this model. Then

$$\Omega_t = B\Sigma_t B' + \Psi, \quad \text{where } \Sigma_t = \text{diag}(\sigma_{it}^2, \dots, \sigma_{Kt}^2).$$

As $B\Sigma_t B'$ is deficient of rank by $N - K$ there exits an $N \times N$ matrix β such that $\beta B = (0 : I_k)$. Consequently, if we write $\beta = (\beta_1', \dots, \beta_N')$, this model allows $N - K$ portfolios $\beta_k' y_t$ to be formed which are homoscedastic. The other portfolios do not have time-varying covariances, just varying variances σ_{kt}^2.

This elegant result suggests forcing σ_{kt}^2 to vary as a GARCH model shocked by past portfolio values $\beta_k' y_t$. Consequently the K factor GARCH(1,1) model becomes:

$$\Omega_t = \Psi^* + \sum_{k=1}^{K} \alpha_k (b_k \beta_k' y_{t-1} y_{t-1}' \beta_k b_k') + \sum_{k=1}^{K} \gamma_k (b_k \beta_k' \Omega_{t-1} \beta_k b_k').$$

Here α_k and γ_k are $N \times N$ matrices. In this model the b_k and β_k are constrained so that $b_k' \beta_j = I(k = j)$ and $\beta_j' \iota = 1$,where ι is the unit vector. Estimation of this type of model is discussed at length by Lin (1992).

1.4.4 Unobserved ARCH

The use of the factor structure seems a real step forward, but the mechanism for driving the σ_{kt}^2 in the factor ARCH model seems quite involved. A simple structure could be obtained by allowing σ_{kt}^2 to be an ARCH process in the unobserved factors f_{it}. This is the suggestion of Diebold and Nerlove (1989) and has been refined by King, Sentana and Wadhwani (1994). The econometrics of this model is a straightforward generalization of the univariate case outlined in section 1.2.

1.4.5 Multivariate SV models

Some multivariate SV models are easy to state. Harvey, Ruiz and Shephard (1994) used quasi-likelihood Kalman filtering techniques on

$$y_{it} = \varepsilon_{it} \exp(h_{it}/2), \quad i = 1, \ldots, N,$$

where

$$\varepsilon_t = (\varepsilon_{1t}, \ldots, \varepsilon_{NT})' \sim NID(0, \Sigma_\varepsilon),$$

in which Σ_ε is a correlation matrix (this model can be viewed as a generalization of the discounted dynamic models of Quintana and West, 1987). They allow $h_t = (h_{it}, \ldots, h_{NT})'$ to follow a multivariate random walk, although more complicated linear dynamics could be handled (Harvey, 1989, Chapter 8). The approach again relies on linearizing (this time with loss of information) by writing $\log y_{it}^2 = h_{it} + \log \varepsilon_{it}^2$. The components of the vector of $\log \varepsilon_{it}^2$ are iid, all with means -1.2704, and a covariance matrix which is a known function of Σ_ε (Harvey, Ruiz and Shephard, 1994, p. 251). Consequently Σ_ε and the parameters indexing the dynamics of h_t can be estimated.

There are two basic points about this model. First, it allows common trends and cycles in volatility by placing reduced rank constraints on h_t, paralleling the work of Harvey and Stock (1988) on the levels of income and consumption. Second, the model is one of changing variances, rather than changing correlation, similar to the Bollerslev (1990) model of constant conditional correlation. Consequently, this model can be empirically successful, but it is of limited interest since it cannot model some theoretically important features of the data. Other work on this model includes Mahieu and Schotman (1994), while Jacquier, Polson and Rossi (1995) look at using an MCMC sampler on this model.

1.4.6 Multivariate factor SV model

Perhaps a more attractive multivariate SV model can be built out of factors. The simplest one-factor model puts

$$\begin{aligned} y_t &= \beta f_t + w_t, \quad w_t \sim NID(0, \Sigma_w) \\ f_t &= \varepsilon_t \exp(h_t/2), \text{ where } h_{t+1} = \gamma_1 h_t + \eta_t, \quad \eta_t \sim NID(o, \sigma_\eta^2). \end{aligned}$$

Here w_t obscures the scaled univariate SV model f_t. Typically Σ_w will be assumed diagonal, perhaps driven by independent SV models. It is similar in spirit to the Diebold and Nerlove (1989) model.

Direct Kalman filtering methods do not seem effective on these models as there is no obvious linearizing transformation. MCMC methods do not suffer this drawback and are explored in Jacquier, Polson and Rossi (1995) and Pitt and Shephard (1995).

1.5 Option pricing with changing volatility

There is now a considerable literature on computing option prices on assets with variances which change over time. Most – for example Taylor (1986, Ch. 9), Hull and White (1987), Scott (1987), Wiggins (1987), Melino and Turnbull (1990), Stein and Stein (1991) and Heston (1993) – use an SV framework (all but Taylor in continuous-time). Recently there has been some work on option pricing based on ARCH-driven models; see Engle and Mustafa (1992), Engle, Hong, Kane and Noh (1993), Engle, Kane and Noh (1993), Heynen, Kemna and Vorst (1994), Noh, Engle and Kane (1994) and Duan (1995). In both cases it is not possible to use the risk-elimination methods discussed earlier for the Black–Scholes constant variance solution, for there is no direct market which trades volatility. However, under the assumption that volatility is uncorrelated with aggregate consumption, progress can be made (Hull and White, 1987, p. 283). In this discussion we will also assume that volatility is uncorrelated with the stock price itself, a less satisfactory assumption except for currencies, although this can be removed.

1.5.1 Stochastic volatility

Hull and White (1987) set up a risk-neutral world and exploit the standard diffusion for $dS = rSdt + \sigma Sdz$, the stock price, and add $dV = \lambda V dt + \theta V dw$, allowing λ to depend on $V = \log \sigma^2$. Then they show a fair European call option price would be

$$\begin{aligned} hw_v &= e^{-rv} \int \max\{S(T+v) - K, 0\} \\ &\quad f\{S(T+v) \mid S(T), \sigma^2(T)\} dS(T+v), \end{aligned}$$

assuming $\sigma^2(T)$ is observable.

This integral has been solved analytically, using characteristic function inversions, in a number of recent papers (Stein and Stein, 1991; Heston 1993). There are, however, substantial benefits from simplification which can be used if simple simulation techniques are to be used to approximate it. If we write σ^2 as the sample path of the volatility, then

$$\log\{S(T+v)/S(T)|\sigma^2\} \sim N(rv - v\overline{\sigma}^2/2, v\overline{\sigma}^2),$$

where

$$\overline{\sigma}^2 = \frac{1}{v}\int_T^{T+v} \sigma^2(t)dt,$$

is the average level of volatility during the option. This implies

$$hw_v = e^{-rv}\iint \max\{S(\tau+v)-K,0\}f\{S(T+v)\mid S(T),\overline{\sigma}^2\}$$
$$dS(T+v)f\{\overline{\sigma}^2 \mid \sigma^2(T)\}d\overline{\sigma}^2$$

The inner of these integrals is the Black–Scholes pricing formula (1.2) replacing σ^2 by $\overline{\sigma}^2$. This leaves the integral in its 'Rao–Blackwellized' form,

$$hw_v = \int bs_v(\overline{\sigma}^2)f\{\overline{\sigma}^2 \mid \sigma^2(T)\}d\overline{\sigma}^2, \qquad (1.29)$$

which is much easier to solve by simulation.

In practice, the diffusions will be discretized, perhaps into the SV models discussed earlier. Then we will need to take into account that $\sigma^2(T)$ is unobserved.

A simple approach to this problem is to draw m MCMC replications from $f(h \mid y)$ to construct a fixed population of initial conditions $h_T^1, \ldots, h_T^m$. Then, sampling with replacement from this population, we can initialize a draw from a whole sequence of future $h_{T+1}^j, \ldots, h_{T+v}^j$. Each sequence draws a mean

$$\overline{\sigma}^{2j} = \frac{1}{v}\sum_{t=1}^{v} \exp(h_{T+t}^j)$$

to provide an estimate of (1.29) $(1/R)\sum_{j=1}^R bs_v(\overline{\sigma}^{2j})$.

As Hull and White (1987) noted, it is possible to use antithetic variables productively (Ripley, 1987, pp. 129–132) for the shocks η_t^j in the autoregression, since h_t^j appear monotonically in the replications of $\overline{\sigma}^{2j}$. This means that if η_t^j are negatively correlated across j then the resulting $\overline{\sigma}^{2j}$ will also be negatively correlated (see Ripley, 1987, Theorem 5.2, p. 129), reducing the variance of the Monte Carlo estimate. An obvious example is to draw double replications based on

$$h_{T+t+1}^{j+} = \gamma_1 h_{T+t}^{j+} + \eta_{T+t} \quad \text{and} \quad h_{T+t+1}^{j-} = \gamma_1 h_{T+t}^{j-} - \eta_{T+t},$$

starting off $h_T^{j+} = h_T^{j-}$ at the same point.

1.5.2 ARCH modelling

It is not so straightforward to carry out option pricing based on ARCH models of the form

$$\log S_{t+1} = r + \log S_t + y_t, \qquad \text{where } y_t \sim \text{GARCH}.$$

The main reason for this is that $\log\{S_{T+v}/S_T|\sigma^2\}$ is no longer normally distributed. The implication of this is that the ARCH option pricing formula has to be estimated by the path-dependent summation

$$arch_v = e^{-rv} \sum_{j=1}^{R} \max\{S_{T+v}^j - K, 0\},$$

where S_{T+v}^j are GARCH simulations into the future starting off using Y_T. This is much less satisfactory, implying R will have to be much higher for this model than for the corresponding SV model as it depends on the sample path of the observations rather than the volatility.

To overcome this difficulty, Noh, Engle and Kane (1994) use a 'plug-in' estimate of future average volatility to deliver

$$arch_v \simeq bs_v(\sigma_{t+1|t}^{(v)}) \quad \text{where } \sigma_{t+1|t}^{(v)} = \frac{1}{v} \sum_{i=1}^{v} Ey_{t+i}^2 | Y_T.$$

Although this is an approximation, based on the expectation of a function being approximately the function of the expectation, Noh, Engle and Kane (1994) provide some evidence that this is a sufficiently good improvement over existing technology to provide a trading profit on at-the-money (where $K = y_T$) Standard and Poor's 500 index European options. Further, from a theoretical viewpoint Cox and Rubinstein (1985, p. 218) show that the Black–Scholes pricing equation is essentially a linear function of the standard deviation around at-the-money prices. Hence the plug-in approximation is likely to be good for at-the-money options, although at other prices it could be poor.

1.5.3 Applying volatility models

Even though it is not possible to predict the way the market is going to move, it may be possible to use the volatility models to trade profitably in the options market. Suppose we believe that in the next v periods the market will be more volatile than do other traders. We should buy both call and put options: if the market goes down we would exercise the put option, if it goes up the call option would be used. Thus this **straddle** is of value in volatile times. If the opposite is true, we can sell a straddle.

This strategy of using straddles potentially allows us economically to value the use of the various volatility models: implied volatility, ARCH and stochastic volatility. The first two of these are compared in Noh, Engle and Kane (1994). I have not seen any work on using straddles to value SV models. Of course in practice this type of trading, particularly the selling of straddles, can be very risky; see the report in the *Financial Times* on the collapse of Barings Bank by Martin (1995).

1.6 Continuous-time models

Continuous-time models play a crucial role in finance and some economic theory, often providing very simple and powerful solutions to difficult problems. An example of this is the Hull and White (1987) generalization of the Black–Scholes option pricing formula to allow the instantaneous variance of returns to change over time. Two obvious questions are:

- Should we use continuous-time (tick-by-tick) data to estimate continuous- and discrete-time models?
- How do we use daily or weekly data to estimate continuous-time models?

I think the answer to the first of these is probably no. The reason for this is that, as with continuous-time models in other fields, continuous-time models are grossly untrue over short periods of time due to institutional factors. An example of this is the large seasonal patterns of volatility which occur during the day (see Foster and Viswanathan, 1990), due to the opening, lunch hour and closing of various markets around the world. These types of feature introduce high-dimensional unstable nuisance features into the modelling process. As they are of little interest, it seems sensible to abstract from them.

This criticism of the use of continuous-time data suggests that the highly impressive work of Nelson (1992; 1995) and Nelson and Foster (1994), who have used the continuous record asymptotics, may not be a particularly fruitful approach to the direct estimation of models (although the asymptotics are useful at indicating the link between different models). This is because that work relies on building ARCH-type models by studying approximations (for example in Nelson, 1990b; 1995) of continuous-time SV models as the time gap between observations falls in a particular way. Thus, at least in theory, there is a belief that the continuous-time data could be used to perform filtering and smoothing.

The second question is the subject of considerable research attention at the moment which we will briefly discuss here.

1.6.1 Direct estimation

There is little work on the direct estimation of continuous-time models using discrete-time data. The reason for this is the nonlinear nature of the volatility models which make the tools developed for linear models (see Bergstrom, 1983; Harvey and Stock, 1985) not directly useful.

Recently there has been an upswing of interest in using generalized method of moments criteria to estimate continuous-time SV models. Leading work on this topic includes that by Duffie and Singleton (1993) and Ho, Perraudin and Sorensen (1993). Given the gross inefficiency of GMM in discrete time, it is doubtful that these approaches will be particularly effective at dealing with continuous-time models.

1.6.2 Indirect inference

A promising approach to using a discrete-time model to perform inference on continuous models has been suggested by Smith (1993) and elaborated by Gourieroux, Monfort and Renault (1993) and Gallant and Tauchen (1995). Gourieroux, Monfort and Renault (1993) call this procedure **indirect inference**.

The idea is to use an approximate model's objective function, $Q(y;\beta)$, such as a likelihood or quasi-likelihood from a discrete-time model, as the basis of inference for a fully parametric continuous-time model, indexed by parameters θ, where the likelihood is difficult to calculate. We write

$$\widehat{\beta} = \arg\max_{\beta} Q(y;\beta) \quad \text{and} \quad \widehat{\beta}^h(\theta) = \arg\max_{\beta} Q(\widetilde{y}^h(\theta);\beta)$$

where $\widetilde{y}^h(\theta)$ is the hth simulation using the continuous-time model with the parameter θ. Then we solve

$$\widehat{\theta} = \arg\max_{\theta} \left\{ \widehat{\beta} - \frac{1}{H}\sum_{h=1}^{H} \widetilde{\beta}^h(\theta) \right\}' \Omega^{-1} \left\{ \widehat{\beta} - \frac{1}{H}\sum_{h=1}^{H} \widetilde{\beta}^h(\theta) \right\},$$

for some choice of Ω. The asymptotic properties of $\widehat{\theta}$ are studied in Gourieroux, Monfort and Renault (1993).

A slight variant of this approach can be found in Gallant and Tauchen (1995), who work with

$$\widehat{\theta} = \arg\max_{\theta} \left\{ \frac{1}{H}\sum_{h=1}^{H} \frac{\partial Q}{\partial \beta'}(\widetilde{y}^h(\theta);\widehat{\beta}) \right\}' \Sigma^{-1} \left\{ \frac{1}{H}\sum_{h=1}^{H} \frac{\partial Q}{\partial \beta}(\widetilde{y}^h(\theta);\widehat{\beta}) \right\}.$$

In the case where the dimension of β equals that of θ, this is equivalent to solving for $\widehat{\theta}$ the equation

$$\frac{1}{H}\sum_{h=1}^{H}\frac{\partial Q}{\partial \beta}(\widetilde{y}^h(\widehat{\theta});\widehat{\beta})=0.$$

Note that the data only play a role through $\widehat{\beta}$.

To illustrate this procedure we will first work with a toy example. Suppose $y_i \sim N(\mu,1)$, and that we use a correctly specified model. Then, setting $x_i^h \sim NID(0,1)$ and writing θ as the mean under estimation, the indirect inference estimator becomes

$$\widehat{\theta}=\arg\max_{\theta}\{\overline{y}-(\theta+\overline{x})\}^2,$$

yielding $\widehat{\theta}=\overline{y}+\overline{x}$. The Gallant and Tauchen (1995) approach yields the same estimator, via

$$\frac{\partial Q}{\partial \mu}=\sum_{h=1}^{H}\sum_{i=1}^{T}(y_i^h-\overline{y})=TH(\theta-\overline{y})+\sum_{h=1}^{H}\sum_{i=1}^{T}x_i^h,\text{ where } y_i^h=\theta+x_i^h.$$

This implies $\widehat{\theta}\sim N\left(\theta,\frac{1}{T}(1+\frac{1}{H})\right)$. For more complicated problems $\widehat{\theta}$ remains consistent for finite H as $T\to\infty$.

The indirect inference method can be used to estimate SV models using an ARCH likelihood. One approach to estimating the continuous-time SV models is to use a discrete-time GARCH(1,1) model as a template. This is convenient as the GARCH model has an analytic likelihood. It is also straightforward to simulate from the continuous-time SV model by taking a very fine discrete-time approximation (e.g. split each day into ten parts). This would give consistent estimation of the continuous-time models. This agenda has recently been followed by Engle and Lee (1994). The efficiency of these types of procedure compared to some likelihood-based methods is still open to question.

1.7 Concluding remarks

I have tried to develop a balanced introduction to the current literature on ARCH and SV models. The work on ARCH models is somewhat more mature than the corresponding SV analysis, but both areas of research are vibrant.

I think it is clear from my survey that the development of univariate discrete-time ARCH is a thoroughly mined area. It would be surprising if there were any really major contributions which could be made to this subject. However, multivariate models are still in their infancy and

I believe the best work on this subject has yet to be written. The key to achieving parsimony must be the combination of economic and time series insight.

SV models are much newer. Their close connection to continuous-time models provides a strong motivation; building an appropriate econometric toolbox to treat them is still an ongoing project. Fast MCMC algorithms need to be developed, while the interplay of good empirical model checking and building has only just started for univariate models. There is much to be done.

The use of continuous-time models in theoretical finance provides a large incentive for econometricians to think about how to make inferences on these models using discrete-time data. This is going to be an extremely active area of research in the next decade.

1.8 Appendix

This Appendix focuses on the computational issues of filtering and smoothing. It provides no direct insight into ARCH or SV models.

1.8.1 Gaussian state-space form

The state-space form

$$\begin{array}{rcl} y_t & = & Z_t\alpha_t + G_t u_t, \quad u_t \sim NID(0, I), \\ \alpha_{t+1} & = & T_t\alpha_t + H_t u_t, \\ \alpha_1|Y_0 & \sim & N(a_{1|0}, P_{1|0}), \end{array}$$

has a prominent role in modern time series (see Hannan and Deistler, 1988; Harvey, 1993b). It provides, by appropriate selection of Z_t, G_t, T_t and H_t, a unified representation of all linear Gaussian time series models. For simplicity we assume that $G_t'H_t = 0$ and we write the non-zero rows of H_t as M_t, $G_tG_t' = \Sigma_{\varepsilon t}$ and $H_tH_t' = \Sigma_{\eta t}$. An example of this is an AR(2) with measurement error. Using an obvious notation,

$$Z_t = \begin{pmatrix} 1 & 0 \end{pmatrix}, G_t = \begin{pmatrix} \sigma_\varepsilon & 0 \end{pmatrix}, T_t = \begin{pmatrix} \phi_1 & 1 \\ \phi_2 & 0 \end{pmatrix},$$

$$H_t = \begin{pmatrix} 0 & \sigma_\eta \\ 0 & 0 \end{pmatrix}, M_t = \begin{pmatrix} 0 & \sigma_\eta \end{pmatrix}.$$

1.8.2 Kalman filter

The Kalman filter (Harvey, 1989) plays a crucial computational role in the analysis of models in state-space form. In particular, if we write

$a_{t|t-1} = E\alpha_t|Y_{t-1}$ and $P_{t|t-1} = \text{MSE}(\alpha_t|Y_{t-1})$, then the Kalman filter computes these quantities recursively for $t = 1, \ldots, T$,

$$\begin{array}{lclllcl} a_{t+1|t} & = & T_t a_{t|t-1} + K_t v_t, & P_{t+1|t} & = & T_t P_{t|t-1} L_t' + \Sigma_{\eta t}, \\ v_t & = & y_t - Z_t a_{t|t-1}, & F_t & = & Z_t P_{t|t-1} Z_t' + \Sigma_{\varepsilon t}, \\ K_t & = & T_t P_{t|t-1} Z_t' F_t^{-1}, & L_t & = & T_t - K_t Z_t. \end{array} \tag{1.30}$$

A by-product of the filter are the innovations v_t, which are the one-step-ahead forecast errors, and their corresponding mean squared errors, F_t. Together they deliver the likelihood, such as (1.17).

1.8.3 Simulation smoother

Traditionally the posterior density of α given Y_T is called the **smoothing density**. In the Gaussian case there are algorithms which compute the mean and covariance terms of this highly multivariate distribution. I call these algorithms **analytic smoothers**. More recently a number of algorithms, labelled **simulation smoothers**, have been proposed for drawing random numbers from $\alpha|Y_T$. This is useful in Gibbs sampling problems. The first algorithms were proposed by Carter and Kohn (1994) and Fruhwirth-Schnatter (1994). We exploit the approach of de Jong and Shephard (1995), which is computationally simpler, more efficient and avoids singularities. This requires that F_t^{-1}, v_t and K_t be stored from the run of the Kalman filter.

Setting $r_T = 0$ and $N_T = 0$, we run for $t = T, \ldots, 1$

$$\begin{array}{lcl} C_t & = & M_t M_t' - M_t H_t' N_t H_t M_t', \\ r_{t-1} & = & Z_t' F_t^{-1} v_t + L_t' r_t - L_t' N_t H_t M_t' C_t^{-1} \kappa_t, \\ \kappa_t & \sim & N(0, C_t), \\ N_{t-1} & = & Z_t' F_t^{-1} Z_t + L_t' N_t L_t + L_t' N_t H_t M_t' C_t^{-1} M_t H_t' N_t L_t, \end{array} \tag{1.31}$$

storing the simulated $M_t u_t$ as $\widehat{M_t u_t} = M_t H_t' r_t + \kappa_t$. It will then be convenient to add to $\widehat{M_t u_t}$ the corresponding zero rows so that we simulate from the whole $H_t u_t$ vector (recall H_t is M_t plus some rows of zeros). We will write this as $\widehat{\eta}_t$.

The end condition $\widehat{\eta}_0$ is calculated by

$$\begin{array}{lcllcl} C_0 & = & P_{1|0} - P_{1|0} N_0 P_{1|0}, & \kappa_0 & \sim & N(0, C_0), \\ \widehat{\eta}_0 & = & P_{1|0} r_0 + \kappa_0. & & & \end{array}$$

The α vector to be simulated via the forward recursion, starting with $\alpha_0 = 0$, is given by

$$\alpha_{t+1} = T_t \alpha_t + \widehat{\eta}_t, \quad t = 0, \ldots, T-1. \tag{1.32}$$

1.8.4 Analytic smoothing

Analytic smoothing is useful in estimating the unobserved log-volatility in the SV models. In particular, the analytic smoother, due to de Jong (1989), computes $a_{t|T} = E\alpha_t|Y_T$ and $P_{t|T} = \text{MSE}(\alpha_t|Y_T)$. It requires that $\alpha_{t+1|T}, P_{t+1|T}, F_t^{-1}, v_t$ and K_t be stored from the Kalman filter. Then, setting $r_T = 0$ and $N_t = 0$, it computes backwards:

$$r_{t-1} = Z_t' F_t^{-1} v_t + L_t' r_t, \quad N_{t-1} = Z_t' F_t^{-1} Z_t + L_t' N_t L_t, t = T, \ldots, 1.$$

Then it records

$$\alpha_{t+1|T} = \alpha_{t|t-1} + P_{t|t-1} r_{t-1} \text{ and } P_{t|T} = P_{t|t-1} - P_{t|t-1} N_{t-1} P_{t|t-1}.$$

1.9 Computing and data sources

There is very little publicly available software to fit ARCH and SV models. The most developed set of programs are those available in TSP and SAS which perform GARCH-M estimation using a normal target. EVIEWS allows some analysis of multivariate GARCH models. The unobserved components time series software STAMP allows a quasi-likelihood analysis of SV models (see Koopman *et al.*, 1995).

All the calculations reported in this paper were performed using my own FORTRAN code compiled using WATCOM 9.5. Throughout I used NAG subroutines to generate random numbers and E04JAF to do standard numerical optimization. I thank Wally Gilks and Peter Rossi for sending me their code to perform random number generation and MCMC analysis of SV models, respectively.

All series reported in this paper are taken from the UK's DATASTREAM. DATASTREAM does not record the exchange rates at weekends (even though there is some very thin trading), and so gives roughly 261 observations a year. The series records the previous day's value if the market is closed during a weekday. Examples of this are Christmas day and bank holidays. These have not been taken out in the analysis presented in this paper.

Acknowledgements

I wish to thank the organizing committee of the Séminaire Européen de Statistique for their invitation to give the lectures on which this paper is based. I am very grateful to Ole Barndorff-Nielsen, David Cox, Eric Ghysels, Andrew Harvey, David Hendry, David Hinkley, Wilfrid Kendall, Lars Korsholm, Aurora Manrique, Mike Pitt, Tina Rydberg, Enrique Sentana and Stephen Taylor for their comments on a previous version of this paper and to Jim Durbin for allowing me to quote his suggestion of finding the mode in the SV problem. I would like to thank the ESRC for their financial support and Maxine Brant for typing the first draft of this paper.

References

Andersen, T. (1995) Return volatility and trading volume: an information flow interpretation of stochastic volatility. *Journal of Finance*. Forthcoming.

Andersen, T. and B. Sorensen (1995) GMM estimation of a stochastic volatility model: a Monte Carlo study. *Journal of Business and Economic Statistics*. Forthcoming.

Baillie, R.T. and T. Bollerslev (1990) A multivariate generalized ARCH approach to modelling risk premium in forward foreign exchange rate markets. *Journal of International Money and Finance*, **9**, 309–324.

Baillie, R.T. and T. Bollerslev (1992) Prediction in dynamic models with time-dependent conditional variances. *Journal of Econometrics*, **52**, 91–113.

Baillie, R.T., T. Bollerslev and H.O. Mikkelsen (1995). Fractionally integrated generalized autoregressive conditional heteroscedasticity. *Journal of Econometrics*. Forthcoming.

Bartholomew, D.J. (1987) *Latent Variable Models and Factor Analysis*. Oxford University Press, New York.

Bera, A.K. and M.L. Higgins (1995) On ARCH models: properties, estimation and testing. In L. Oxley, D.A.R. George, C.J. Roberts, and S. Sayer (eds), *Surveys in Econometrics*. Oxford: Blackwell. Reprinted from *Journal of Economic Surveys*.

Bergstrom, A.R. (1983) Gaussian estimation of structural parameters in higher order continuous time dynamic models. *Econometrica*, **51**, 117–151.

Berzuini, C., N.G. Best, W.R. Gilks, and C. Larizza (1994) Dynamic graphical models and Markov chain Monte Carlo methods. Unpublished paper: MRC Biostatistics Unit, Cambridge.

Beveridge, S. and C.R. Nelson (1981) A new approach to decomposition of economic time series into permanent and transitory components with particular attention to measurement of the business cycle. *Journal of Monetary Economics*, **7**, 151–174.

Black, F. (1976) Studies of stock price volatility changes. *Proceedings of the American Statistical Association, Business and Economic Statistics Section*, 177–181.

Black, F. and M. Scholes (1973) The pricing of options and corporate liabilities. *Journal of Political Economy*, **81**, 637–654.

Bollerslev, T. (1986) Generalised autoregressive conditional heteroscedasticity. *Journal of Econometrics*, **51**, 307–327.

Bollerslev, T. (1987) A conditional heteroscedastic time series model for speculative prices and rates of return. *Review of Economics and Statistics*, **69**, 542–547.

Bollerslev, T. (1990) Modelling the coherence in short-run nominal exchange rates: a multivariate generalized ARCH approach. *Review of Economics and Statistics*, **72**, 498–505.

Bollerslev, T., R.Y. Chou, and K.F. Kroner (1992) ARCH modeling in finance: A review of the theory and empirical evidence. *Journal of Econometrics*, **52**, 5–59.

Bollerslev, T. and R.F. Engle (1993) Common persistence in conditional variances. *Econometrica*, **61**, 167–186.

Bollerslev, T., R.F. Engle and D.B. Nelson (1995) ARCH models. In R.F. Engle and D. McFadden (eds), *The Handbook of Econometrics, Volume 4*. North-Holland, Amsterdam. Forthcoming.

Bollerslev, T., R.F. Engle and J.M. Wooldridge (1988) A capital asset pricing model with time varying covariances. *Journal of Political Economy*, **96**, 116–131.

Bollerslev, T. and J.M. Wooldridge (1992) Quasi maximum likelihood estimation and inference in dynamic models with time varying covariances. *Econometric Reviews*, **11**, 143–172.

Bougerol, P. and N. Picard (1992) Stationarity of GARCH processes and of some non-negative time series. *Journal of Econometrics*, **52**, 115–128.

Breidt, F.J. and A.L. Carriquiry (1995) Improved quasi-maximum likelihood estimation for stochastic volatility models. Unpublished paper: Department of Statistics, Iowa State University.

Breidt, F.J., N. Crato and P. de Lima (1993) Modelling long-memory stochastic volatility. Unpublished paper: Statistics Department, Iowa State University.

Bresnahan, T.F. (1981) Departures from marginal-cost pricing in the American automobile industry, estimates from 1977–1978. *Journal of Econometrics*, **17**, 201–227.

Brock, W., J. Lakonishok and B. LeBaron (1992) Simple technical trading rules and the stochastic properties of stock returns. *Journal of Finance*, **47**, 1731–1764.

Campbell, J.Y. and L. Hentschel (1992) No news is good news: an asymmetric model of changing volatility in stock returns. *Journal of Financial Economics*, **31**, 281–318.

Carlin, B.P., N.G. Polson and D. Stoffer (1992) A Monte Carlo approach to nonnormal and nonlinear state-space modelling. *Journal of the American Statistical Association*, **87**, 493–500.

Carter, C.K. and R. Kohn (1993) On the applications of Markov chain Monte Carlo methods to linear state space models. *Proceedings of the American Statistical Association, Business and Economic Statistics Section*, 131–136.

Carter, C.K. and R. Kohn (1994) On Gibbs sampling for state space models. *Biometrika*, **81**, 541–553.

Chan, K.S. and J. Ledolter (1995) Monte Carlo EM estimation for time series models involving counts. *Journal of the American Statistical Association*, **89**, 242–252.

Chesney, M. and L.O. Scott (1989) Pricing European options: a comparison of the modified Black–Scholes model and a random variance model. *Journal of Financial and Qualitative Analysis*, **24**, 267–284.

Clark, P.K. (1973) A subordinated stochastic process model with fixed variance for speculative prices. *Econometrica*, **41**, 135–156.

Cox, D.R. (1981) Statistical analysis of time series: some recent developments. *Scandanavian Journal of Statistics*, **8**, 93–115.

Cox, J.C. and M. Rubinstein (1985) *Options Markets*. Prentice Hall, Englewood Cliffs, NJ.

Cramér, H. (1946) *Mathematical Methods of Statistics*. Princeton University Press, Princeton, NJ.

Danielsson, J. (1994) Stochastic volatility in asset prices: estimation with simulated maximum likelihood. *Journal of Econometrics*, **61**, 375–400.

Danielsson, J. and J.F. Richard (1993) Accelerated Gaussian importance sampler with application to dynamic latent variable models. *Journal of Applied Econometrics*, **8**, S153–S174.

Davidian, M. and R.J. Carroll (1987) Variance function estimation. *Journal of the American Statistical Association*, **82**, 1079–1091.

de Jong, P. (1989) Smoothing and interpolation with the state space model. *Journal of the American Statistical Association*, **84**, 1085–1088.

de Jong, P. and N. Shephard (1995) The simulation smoother for time series models. *Biometrika*, **82**, 339–350.

Demos, A. and E. Sentana (1994) Testing for GARCH effects: a one sided approach. Unpublished: CEMFI, Madrid.

Diebold, F.X. and J.A. Lopez (1995) ARCH models. In K. Hoover (ed.), *Macroeconomics: Developments, Tensions and Prospects*.

Diebold, F.X. and M. Nerlove (1989) The dynamics of exchange rate volatility: a multivariate latent factor ARCH model. *Journal of Applied Econometrics*, **4**, 1–21.

Ding, Z., C.W.J. Granger, and R.F. Engle (1993) A long memory property of stock market returns and a new model. *Journal of Empirical Finance*, **1**, 83–106.

Drost, F.C. and T.E. Nijman (1993) Temporal aggregation of GARCH processes. *Econometrica*, **61**, 909–927.

Drost, F.C. and B. Werker (1993) Closing the GARCH gap: continuous time GARCH modelling. Unpublished paper: CentER, Tilburg University.

Duan, J.C. (1995) The GARCH option pricing model. *Mathematical Finance*, **6**.

Duffie, D. and K.J. Singleton (1993) Simulated moments estimation of Markov models of asset process. *Econometrica*, **61**, 929–952.

Dunsmuir, W. (1979) A central limit theorem for parameter estimation in stationary vector time series and its applications to models for a signal observed with noise. *Annals of Statistics*, **7**, 490–506.

Durbin, J. (1992) Personal correspondence to Andrew Harvey and Neil Shephard.

Durbin, J. (1996) *Time Series Analysis Based on State Space Modelling for Gaussian and Non-Gaussian Observations*. Oxford: Oxford University Press. RSS Lecture Notes Series.

Durbin, J. and S.J. Koopman (1992) Filtering, smoothing and estimation for time series models when the observations come from exponential family distributions. Unpublished paper: Department of Statistics, London School of Economics.

Engle, R.F. (1982) Autoregressive conditional heteroscedasticity with estimates of the variance of the United Kingdom inflation. *Econometrica*, **50**, 987–1007.

Engle, R.F. (1995) *ARCH*. Oxford: Oxford University Press.

Engle, R.F. and T. Bollerslev (1986) Modelling the persistence of conditional variances. *Econometric Reviews*, **5**, 1–50, 81–87.

Engle, R.F. and G. Gonzalez-Rivera (1991) Semiparametric ARCH models. *J. Economics and Business Statist.*, **9**, 345–359.

Engle, R.F., D.F. Hendry and D. Trumble (1985) Small-sample properties of ARCH estimators and tests. *Canadian Journal of Economics*, **18**, 66–93.

Engle, R.F., C.-H. Hong, A. Kane and J. Noh (1993) Arbitrage valuation of variance forecasts with simulated options. *Advances in Futures and Options Research*, **6**, 393–415.

Engle, R.F., A. Kane and J. Noh (1993) Option-index pricing with stochastic volatility and the value of accurate variance forecasts. Unpublished paper: Department of Economics, University of California at San Diego.

Engle, R.F. and K.F. Kroner (1995) Multivariate simultaneous generalized ARCH. *Econometric Theory*, **11**, 122–150.

Engle, R.F. and G.G.J. Lee (1992) A permanent and transitory component model of stock return volatility. Unpublished paper: Department of Economics, University of California at San Diego.

Engle, R.F. and G.G.J. Lee (1994) Estimating diffusion models of stochastic volatility. Unpublished paper: Department of Economics, University of California at San Diego.

Engle, R.F., D.M. Lilien and R.P. Robins (1987) Estimating time-varying risk premium in the term structure: the ARCH-M model. *Econometrica*, **55**, 391–407.

Engle, R.F. and C. Mustafa (1992) Implied ARCH models for options prices. *Journal of Econometrics*, **52**, 289–311.

Engle, R.F. and V.K. Ng (1993) Measuring and testing the impact of news on volatility. *Journal of Finance*, **48**, 1749–1801.

Engle, R.F., V.K. Ng and M. Rothschild (1990) Asset pricing with a factor ARCH covariance structure: empirical estimates for Treasury bills. *Journal of Econometrics*, **45**, 213–238.

Engle, R.F. and J.R. Russell (1994) Forecasting transaction rates: the autogressive conditional duration model. Unpublished paper: Department of Economics, University of California at San Diego.

Evans, M., N. Hastings, and B. Peacock (1993) *Statistical Distributions* (2nd edn) John Wiley and Sons, New York.

Fahrmeir, L. (1992) Posterior mode estimation by extended Kalman filtering for multivariate dynamic generalised linear models. *Journal of the American Statistical Association*, **87**, 501–509.

Fama, E. (1965) The behaviour of stock market prices. *Journal of Business*, **38**, 34–105.

Foster, F.D. and S. Viswanathan (1990) A theory of the interday variations in volume, variance and trading costs in securities markets. *Rev. Financial Studies*, **3**, 593–624.

Fruhwirth-Schnatter, S. (1994) Data augmentation and dynamic linear models. *Journal of Time Series Analysis*, **15**, 183–202.

Fuller, W. A. (1996) *Introduction to Time Series* (2nd edn). John Wiley, New York. Forthcoming.

Gallant, A.R., D. Hsieh and G. Tauchen (1991) On fitting recalcitrant series: the pound/dollar exchange rate, 1974–83. In W.A. Barnett, J. Powell and G. Tauchen (eds), *Nonparametric and Semiparametric Methods in Economics and Statistics*. Cambridge University Press, Cambridge.

Gallant, A.R., D. Hsieh and G. Tauchen (1994) Estimation of stochastic volatility models with diagnostics. Unpublished paper: Department of Economics, Duke University.

Gallant, A.R. and G. Tauchen (1995) Which moments to match. *Econometric Theory*, **11**. Forthcoming.

Geweke, J. (1986) Modelling the persistence of conditional variances: a comment. *Econometric Reviews*, **5**, 57–61.

Geweke, J. (1989) Exact predictive densities in linear models with ARCH disturbances. *Journal of Econometrics*, **44**, 307–325.

Geweke, J. (1994) Bayesian comparison of econometric models. Unpublished paper: Federal Reserve Bank of Minneapolis.

Gilks, W.R. and P. Wild (1992) Adaptive rejection sampling for Gibbs sampling. *Applied Statistics*, **41**, 337–348.

Glosten, L.R., R. Jagannathan and D. Runkle (1993) Relationship between the expected value and the volatility of the nominal excess return on stocks. *Journal of Finance*, **48**, 1779–1802.

Gourieroux, C. and A. Monfort (1992) Qualitative threshold ARCH models. *Journal of Econometrics*, **52**, 159–200.

Gourieroux, C., A. Monfort and E. Renault (1993) Indirect inference. *Journal of Applied Econometrics*, **8**, S85–S118.

Granger, C. W.J. and R. Joyeux (1980) An introduction to long memory time series models and fractional differencing. *Journal of Time Series Analysis*, **1**, 15–39.

Hajivassiliou, V. and D. McFadden (1990) The method of simulated scores, with application to models of external debt crises. Cowles Foundation Discussion Paper, 967.

Hamilton, J. (1994) *Time Series Analysis*. Princeton University Press, Princeton, NJ.

Hannan, E.J. and M. Deistler (1988) *The Statistical Theory of Linear Systems*. John Wiley, New York.

Hansen, L.P. (1982) Large sample properties of generalized method of moments estimators. *Econometrica*, **50**, 1029–54.

Hansen, L.P. and K.J. Singleton (1983) Stochastic consumption, risk aversion and the temporal behavour of asset returns. *Journal of Political Economy*, **91**, 249–265.

Harrison, J. and C.F. Stevens (1976) Bayesian forecasting (with discussion) *Journal of the Royal Statistical Society, Series B*, **38**, 205–247.

Harvey, A.C. (1989) *Forecasting, Structural Time Series Models and the Kalman Filter*. Cambridge University Press, Cambridge.

Harvey, A.C. (1993a) Long memory and stochastic volatility. Submitted.

Harvey, A.C. (1993b) *Time Series Models* (2nd edn) Philip Allan, New York.

Harvey, A.C. and C.Fernandes (1989) Time series models for count data or qualitative observations. *Journal of Business and Economic Statistics*, **7**, 407–417.

Harvey, A.C., E. Ruiz and E. Sentana (1992) Unobserved component time series models with ARCH disturbances. *Journal of Econometrics*, **52**, 129–158.

Harvey, A.C., E. Ruiz and N. Shephard (1994) Multivariate stochastic variance models. *Review of Economic Studies*, **61**, 247–264.

Harvey, A.C. and N. Shephard (1993a) Estimation and testing of stochastic variance models. STICERD Econometrics Discussion Paper, London School of Economics.

Harvey, A.C. and N. Shephard (1993b) The estimation of an asymmetric stochastic volatility model for asset returns. Unpublished paper: Statistics Department, London School of Economics.

Harvey, A.C. and J.H. Stock (1985) The estimation of higher order continuous time autoregressive models. *Econometric Theory*, **1**, 97–112.

Harvey, A.C. and J.H. Stock (1988) Continuous time autoregressive models with common stochastic trends. *Journal of Economic Dynamics and Control*, **12**, 365–384.

Heston, S. L. (1993) A closed-form solution for options with stochastic volatility, with applications to bond and currency options. *Review of Financial Studies*, **6**, 327–343.

Heynen, R., A. Kemna and T. Vorst (1994) Analysis of the term structure of implied volatility. *Journal of Financial Quantitative Analysis*, **29**, 31–56.

Higgins, M.L. and A.K. Bera (1992) A class of nonlinear ARCH models. *International Economic Review*, **33**, 137–158.

Ho, M.S., W.R.M. Perraudin, and B.E. Sorensen (1993) Multivariate tests of a continuous time equilibrium arbitrage pricing theory with conditional heteroscedasticity and jumps. Unpublished paper: Department of Applied Economics, Cambridge University.

Hong, P.Y. (1991) The autocorrelation structure for the GARCH-M process. *Economic Letters*, **37**, 129–132.

Hosking, J.R.M. (1981) Fractional differencing. *Biometrika*, **68**, 165–176.

Hull, J. (1993) *Options, Futures, and Other Derivative Securities* (2nd edn). Prentice Hall International Editions, Englewood Cliffs, NJ.

Hull, J. and A. White (1987) The pricing of options on assets with stochastic volatilities. *Journal of Finance*, **42**, 281–300.

Ingersoll, J. E. (1987) *Theory of Financial Decision Making*. Rowman & Littlefield, Savage, MD.

Jacquier, E., N.G. Polson, and P.E. Rossi (1994) Bayesian analysis of stochastic volatility models (with discussion). *Journal of Business and Economic Statistics*, **12**, 371–417.

Jacquier, E., N.G. Polson, and P.E. Rossi (1995) Models and prior distributions for multivariate stochastic volatility. Unpublished paper: Graduate School of Business, University of Chicago.

Kalman, R.E. (1960) A new approach to linear filtering and prediction problems. *Journal of Basic Engineering, Transactions ASMA, Series D*, **82**, 35–45.

Kim, S. and N. Shephard (1994) Stochastic volatility: Optimal likelihood inference and comparison with ARCH models. Unpublished paper: Nuffield College, Oxford.

King, M., E. Sentana and S. Wadhwani (1994) Volatility and links between national stock markets. *Econometrica*, **62**, 901–933.

Kitagawa, G. and W. Gersch (1984) A smoothness prior — state space modeling of time series with trend and seasonality. *Journal of the American Statistical Association*, **79**, 378–89.

Koenker, R., P. Ng and S. Portnoy (1994) Quantile smoothing splines. *Biometrika*, **81**, 673–80.

Koopman, S.J., A.C. Harvey, J.A. Doornik, and N. Shephard (1995) *STAMP 5.0: Structural Time Series Analyser, Modeller and Predictor*. Chapman & Hall, London.

Kraft, D.F. and R.F. Engle (1982) Autoregressive conditional heteroscedasticity in multiple time series models. Unpublished paper: Department of Economics, University of California at San Diego.

Lee, J.H.H. and M.L. King (1993) A locally most mean powerful based score test for ARCH and GARCH regression disturbances. *Journal of Business and Economic Statistics*, **11**, 17–27.

Lee, S.-W. and B.E. Hansen (1994) Asymptotic theory for the GARCH(1,1) quasi-maximum likelihood estimator. *Econometric Theory*, **10**, 29–52.

Lin, W.-L. (1992) Alternative estimators for factor GARCH models — a Monte Carlo comparison. *Journal of Applied Econometrics*, **7**, 259–279.

Linton, O. (1993) Adaptive estimation in ARCH models. *Econometric Theory*, **9**, 539–569.

Liu, J., W.H. Wong and A. Kong (1994) Covariance structure of the Gibbs sampler with applications to the comparison of estimators and augmentation schemes. *Biometrika*, **81**, 27–40.

Lumsdaine, R.L. (1991) Asymptotic properties of the quasi-maximum likelihood estimator in the GARCH(1,1) and IGARCH(1,1) models. Unpublished paper: Department of Economics, Princeton University.

Mahieu, R. and P. Schotman (1994) Stochastic volatility and the distribution of exchange rate news. Unpublished paper: Department of Finance, University of Limburg.

Mandelbrot, B. (1963) The variation of certain speculative prices. *Journal of Business*, **36**, 394–419.

Martin, P. (1995) Blunders that bust the bank. *Financial Times, 23 March*, p. 24.

McCullagh, P. and J. A. Nelder (1989) *Generalized Linear Models* (2nd edn). Chapman & Hall, London.

Melino, A. and S.M. Turnbull (1990) Pricing foreign currency options with stochastic volatility. *Journal of Econometrics*, **45**, 239–265.

Nelson, D.B. (1990a) Stationarity and persistence in the GARCH(1,1) model. *Econometric Theory*, **6**, 318–334.

Nelson, D.B. (1990b) ARCH models as diffusion approximations. *J. Econometrics*, **45**, 7–38.

Nelson, D.B. (1991) Conditional heteroscedasticity in asset pricing: a new approach. *Econometrica*, **59**, 347–370.

Nelson, D.B. (1992) Filtering and forecasting with misspecified ARCH models I: Getting the right variance with the wrong model. *Journal of Econometrics*, **52**, 61–90.

Nelson, D.B. (1994) Asymptotically optimal smoothing with ARCH models. *Econometrica*, **63**. Forthcoming.

Nelson, D.B. and D.P. Foster (1994) Asymptotic filtering theory for univariate ARCH models. *Econometrica*, **62**, 1–41.

Nijman, T.E. and E. Sentana (1993) Marginalization and contemporaneous aggregation in multivariate GARCH processes. Discussion paper, CentER, Tilburg University.

Noh, J., R.F. Engle, and A. Kane (1994) Forecasting volatility and option pricing of the S&P 500 index. *Journal of Derivatives*, 17–30.

Pagan, A.R. and G.W. Schwert (1990) Alternative models for conditional stock volatility. *Journal of Econometrics*, **45**, 267–290.

Phillips, P. C.B. and S. Durlauf (1986) Multiple time series regression with integrated processes. *Review of Economic Studies*, **53**, 473–495.

Pitt, M. and N. Shephard (1995) Parameter-driven exponential family models. Unpublished paper: Nuffield College, Oxford.

Poon, S. and S.J. Taylor (1992) Stock returns and volatility: an empirical study of the UK stock market. *Journal of Banking and Finance*, **16**, 37–59.

Poterba, J. and L. Summers (1986) The persistence of volatility and stock market fluctuations. *American Economic Review*, **76**, 1124–1141.

Qian, W. and D.M. Titterington (1991) Estimation of parameters in hidden Markov Chain models. *Philosophical Transactions of the Royal Society of London, Series A*, **337**, 407–428.

Quintana, J.M. and M. West (1987) An analysis of international exchange rates using multivariate DLM's. *The Statistician*, **36**, 275–281.

Ripley, B.D. (1987) *Stochastic Simulation*. Wiley, New York.

Ross, S.A. (1976) The arbitrage theory of capital asset pricing. *Journal of Economic Theory*, **13**, 341–360.

Rothenberg, T.J. (1973) *Efficient Estimation with A Priori Information*. Yale University Press, New Haven, CT.

Ruud, P. (1991) Extensions of estimation methods using the EM algorithm. *Journal of Econometrics*, **49**, 305–341.

Schwert, G. W. (1989) Why does stock market volatility change over time? *Journal of Finance*, **44**, 1115–1153.

Scott, L. (1987) Options pricing when the variance changes randomly: theory, estimation and an application. *Journal of Financial and Quantitative Analysis*, **22**, 419–438.

Scott, L. (1991) Random-variance option pricing. *Advances in Future and Options Research*, **5**, 113–135.

Sentana, E. (1991) Quadratic ARCH models: a potential re-intepretation of ARCH models. Unpublished paper: CEMFI, Madrid.

Shephard, N. (1993) Fitting non-linear time series models, with applications to stochastic variance models. *Journal of Applied Econometrics*, **8**, S135–S152.

Shephard, N. (1994a) Local scale model: state space alternative to integrated GARCH processes. *Journal of Econometrics*, **60**, 181–202.

Shephard, N. (1994b) Partial non-Gaussian state space. *Biometrika*, **81**, 115–131.

Smith, A.A. (1993) Estimating nonlinear time series models using simulated vector autoregressions. *Journal of Applied Econometrics*, **8**, S63–S84.

Smith, A. F.M. and G. Roberts (1993) Bayesian computations via the Gibbs sampler and related Markov chain Monte Carlo methods. *Journal of the Royal Statistical Society, Series B*, **55**, 3–23.

Smith, J. Q. (1985) Diagnostic checks of non-standard time series models. *Journal of Forecasting*, **4**, 283–291.

Smith, R.L. and J.E. Miller (1986) A non-Gaussian state space model and application to prediction records. *Journal of the Royal Statistical Society, Series B*, **48**, 79–88.

Steigerwald, D. (1991) Efficient estimation of models with conditional heteroscedasticity. Unpublished paper: Department of Economics, University of California at Santa Barbara.

Stein, E.M. and J. Stein (1991) Stock price distributions with stochastic volatility: an analytic approach. *Review of Financial Studies*, **4**, 727–752.

Tauchen, G. and M. Pitts (1983) The price variability volume relationship on speculative markets. *Econometrica*, **51**, 485–505.

Taylor, S.J. (1986) *Modelling Financial Time Series*. John Wiley, Chichester.

Taylor, S.J. (1994) Modelling stochastic volatility. *Mathematical Finance*, **4**, 183–204.

Uhlig, H. (1992) Bayesian vector autoregressions with time varying error covariances. Unpublished paper: Department of Economics, Princeton University.

Wei, G. C.G. and M.A. Tanner (1990) A Monte Carlo implementation of the EM algorithm and the poor man's data augmentation algorithms. *Journal of the American Statistical Association*, **85**, 699–704.

Weiss, A.A. (1986) Asymptotic theory for ARCH models: estimation and testing. *Econometric Theory*, **2**, 107–131.

West, M. and J. Harrison (1989) *Bayesian Forecasting and Dynamic Models*. Springer-Verlag, New York.

White, H. (1982) Maximum likelihood estimation of misspecified models. *Econometrica*, **50**, 1–25.

White, H. (1994) *Estimation, Inference and Specification Analysis*. Cambridge University Press, Cambridge.

Whittle, P. (1991) Likelihood and cost as path integrals. *Journal of the Royal Statistical Society, Series B*, **53**, 505–538.

Wiggins, J.B. (1987) Option values under stochastic volatilities. *Journal of Financial Economics*, **19**, 351–372.

Wild, P. and W.R. Gilks (1993) AS 287: Adaptive rejection sampling from log-concave density functions. *Applied Statistics*, **42**, 701–709.

Xu, X. and S.J. Taylor (1994) The term structure of volatility implied by foreign exchange options. *Journal of Financial and Quantitative Analysis*, **29**, 57–74.

Zakoian, J.-M. (1990) Threshold heteroscedastic models. Unpublished paper: CREST, INSEE.

Zeger, S.L. and B. Qaqish (1988) Markov regression models for time series, a quasi likelihood approach. *Biometrics*, **44**, 1019–1032.

CHAPTER 2

Likelihood-based inference for cointegration of some nonstationary time series

Søren Johansen

2.1 Introduction

We consider likelihood inference for a class of nonstationary time series and to motivate the discussion we consider some simple economic applications and show how the analysis of the statistical model helps in gaining insight and understanding of economic phenomena. This section contains the basic definitions of integrated variables, cointegration and common trends, and discusses by means of examples the formulation of models and processes in terms of common trends in the moving average representation and of disequilibrium errors in the error correction form of an autoregressive model.

The notion of cointegration has become one of the more important concepts in time series econometrics since the papers by Granger (1983) and Engle and Granger (1987). It has found widespread applications in the analysis of economic data as published in the econometric literature. The special issues of the *Oxford Bulletin of Economics and Statistics*, **54** (3) and *Journal of Policy Modelling*, **14** (3,4), both of 1992, contain many papers where the method is applied and extended. The book of readings by Engle and Granger (1991) contains a collection of papers that have been important for the development of the topic; see also Hargreaves (1994). Many textbooks contain the basic aspects of cointegration: see, for instance, Cuthbertson, Hall and Taylor (1992), Lütkepohl (1993) or Hamilton (1994). The book by Banerjee *et al.* (1993) is a systematic treatment of the subject.

Johansen (1995b) gives a detailed account of the various cointegration models that can be defined as sub-models of the vector autoregressive model. This includes a discussion of the mathematical properties of the processes generated by the models, as well as a discussion of their interpretation. Finally, the book presents a statistical analysis based on the Gaussian likelihood function and the necessary asymptotic theory. The present paper is a summary of the methodology and the results in the book.

In this section we give the basic definitions and a discussion of the concepts, and we start by defining the class of stationary and nonstationary processes we want to investigate. Let ϵ_t denote a doubly infinite sequence of p-dimensional i.i.d. stochastic variables with mean zero and finite variance. From these we construct linear ergodic processes defined by $Y_t = \sum_{i=0}^{\infty} C_i \epsilon_{t-i}$, where the coefficient matrices C_i decrease exponentially fast, so that the series converges almost surely. This implies that the power series

$$C(z) = \sum_{i=0}^{\infty} C_i z^i$$

is convergent for $|z| < 1 + \delta$, for some $\delta > 0$. For the analysis of the likelihood function we need a further condition that the ϵ's are Gaussian. For the asymptotic analysis, however, this condition is not needed, and we only need conditions under which the central limit theorem holds, and for which we get convergence to certain stochastic integrals (section 2.7). In the following we define the concept of $I(0)$ and $I(1)$. The purpose is to define a class of nonstationary processes, $I(1)$, which become stationary after differencing, and a class of stationary processes, $I(0)$, which become nonstationary when summed, thus mimicking the relation between a random walk and its increments.

Definition 1 A linear p-dimensional process $Y_t = \sum_{i=0}^{\infty} C_i \epsilon_{t-i}$ is called **integrated of order zero,** $I(0)$, if $\sum_{i=0}^{\infty} C_i \neq 0$.

Example 1 Consider the stationary univariate autoregressive process which satisfies $Y_t = \rho Y_{t-1} + \epsilon_t$ with $|\rho| < 1$. This is clearly a linear process since $Y_t = \sum_{i=0}^{\infty} \rho^i \epsilon_{t-i}$, and since $\sum_{i=0}^{\infty} C_i = (1-\rho)^{-1} \neq 0$ it follows that it is also $I(0)$. The reason why we want this condition is that the cumulated Y_t's satisfy

$$\sum_{i=1}^{t} Y_i = \left(\sum_{i=0}^{\infty} \rho^i\right) \sum_{i=1}^{t} \epsilon_i - \frac{\rho}{1-\rho} \sum_{i=0}^{\infty} \rho^i (\epsilon_{t-i} - \epsilon_{-i}),$$

which shows that it is the condition $\sum_{i=0}^{\infty} \rho^i \neq 0$ that guarantees that the cumulated process is nonstationary.

The process $X_t = \Delta Y_t = Y_t - Y_{t-1}$, however, is stationary, but not

$I(0)$ since the coefficients sum to zero. If this process is summed we get

$$\sum_{i=1}^{t} X_i = \sum_{i=1}^{t} \Delta Y_i = Y_t - Y_0,$$

which is nonstationary although asymptotically stationary. By insisting on the condition that the sum of the coefficients of the linear process is non-zero, we make sure that the nonstationarity in its cumulated values is of the type we want to describe.

The process composed of both Y_t and ΔY_t, however, is an $I(0)$ process, since the sum of its coefficient matrices is non-zero.

Using the concept of $I(0)$ we now define the main concept for the analysis of cointegration, namely integration of order 1, $I(1)$.

Definition 2 A p-dimensional stochastic process X_t is called **integrated of order 1**, $I(1)$, if ΔX_t is $I(0)$.

The simplest example of an $I(1)$ process is a random walk, but any process of the form

$$X_t = C \sum_{i=1}^{t} \epsilon_i + \sum_{i=0}^{\infty} C_i \epsilon_{t-i},$$

is also an $I(1)$ process, at least if $C \neq 0$. Note that an $I(1)$ process is nonstationary, but that the nonstationarity can be removed by differencing.

Example 2 We define a three-dimensional process by

$$\begin{aligned} X_{1t} &= \textstyle\sum_{i=1}^{t} \epsilon_{1i} + \epsilon_{2t}, \\ X_{2t} &= \textstyle\frac{1}{2} \sum_{i=1}^{t} \epsilon_{1i} + \epsilon_{3t}, \quad t = 0, 1, ..., T, \\ X_{3t} &= \epsilon_{4t}. \end{aligned}$$

It is seen that X_t is nonstationary and that ΔX_t is stationary and $I(0)$. Thus X_t is an $I(1)$ process, which can be made stationary by differencing. It is also seen that $X_{1t} - 2X_{2t}$ is stationary, and we say that X_t is cointegrated with $(1, -2, 0)$ as a cointegrating vector, and the process $\sum_{i=1}^{t} \epsilon_{1i}$ is called a **common stochastic trend**. Thus stationarity can be achieved either by taking differences or by taking suitable linear combinations.

This example illustrates the definition of cointegration:

Definition 3 If X_t is integrated of order 1 but some linear combination, $\beta' X_t, \beta \neq 0$, can be made stationary by a suitable choice of $\beta' X_0$, then X_t is called **cointegrated** and β is the **cointegrating vector**. The number of linearly independent cointegrating vectors is called the **cointegrating rank**, and the space spanned by the cointegrating vectors is the **cointegration space**.

The definition of $I(1)$ is in terms of differences, and nothing is said about the levels of the process. Thus one cannot expect in general that anything can be said about the linear combinations of the levels $\beta' X_t$.Thus if the process defined in Example 2 started with initial values X_{10}, X_{20}, and X_{30}, the linear combination $X_{1t} - 2X_{2t}$ would not be stationary unless the initial value $X_{10} - 2X_{20}$ were given its invariant distribution.

A similar remark holds for the process considered in Example 1, given by

$$Y_t = \rho Y_{t-1} + \epsilon_t, \; t = 1, \ldots, T,$$

which determines Y_t as a function of Y_0 and the ϵ's,

$$Y_t = \rho^t Y_0 + \sum_{i=0}^{t-1} \rho^i \epsilon_{t-i}.$$

This can be made stationary if we choose Y_0 with the right initial distribution, that is, $N(0, \sigma^2/(1-\rho^2))$.

Note that the definition of $I(1)$ is invariant under non-singular linear transformations in the sense that if X_t is $I(1)$ and A is of full rank then AX_t is also $I(1)$. If β is a cointegrating vector for X_t then $A^{-1\prime}\beta$ is a cointegrating vector for AX_t.

The idea behind cointegration is that sometimes the lack of stationarity of a multidimensional process is caused by common stochastic trends, which can be eliminated by taking suitable linear combinations of the process, thereby making the linear combination stationary.

In economics and other applications of statistics autoregressive processes have long been applied to describe stationary phenomena, and the idea of explaining the process by its past values has been very useful for prediction. If, however, we want to find relations between simultaneous values of the variables in order to understand the interactions of the economy we would get a lot more information by relating the value of a variable to the value of other variables at the same time point rather than relating it to its own past. We can say that if we want to discuss relations between variables then we should take combinations of simultaneous values, and if we want to discuss dynamic development of the variables we should investigate the dependence on the past.

The reason why cointegration has been so popular in econometrics is that classical macro-economic models are often formulated as simultaneous linear relations between variables following the Cowles Commission tradition. The theory of such equations was developed for stationary processes despite the fact that many (or even most) economic variables are nonstationary. If we think of the classical economic relations as long-run relations we can easily imagine that such relations can be stationary even if the variables themselves are nonstationary. Cointegration is the mathematical formulation of this phenomenon.

In the autoregressive framework we next discuss the so-called **error correction model**. The concept can be traced back to Phillips (1954) who used ideas from engineering to formulate continuous-time models, and to Sargan (1964) who used the ideas to formulate models for discrete-time data. The simplest example of such a model which still illustrates the main point is as follows.

Example 3 We define processes by the autoregressive model

$$\begin{aligned}\Delta X_{1t} &= -\alpha_1 (X_{1,t-1} - 2X_{2,t-1}) + \epsilon_{1t},\\ \Delta X_{2t} &= \epsilon_{2t}, \quad t = 1, \ldots, T.\end{aligned}$$

This stochastic difference equation models the changes in X_{1t} at time t as reacting through the adjustment coefficient α_1 to a disequilibrium error $X_1 - 2X_2$ at time $t-1$. It is not difficult to see that for $0 < \alpha_1 < 2$ the model defines X_t as nonstationary, but that ΔX_t is $I(0)$, and also $X_{1t} - 2X_{2t}$ is stationary if the initial value $X_{10} - 2X_{20}$ is correctly chosen.

In fact we can solve the equations for X_t as a function of the initial values X_0 and the disturbances $\epsilon_1, \ldots, \epsilon_T$ and find, since $X_{1t} - 2X_{2t}$ is an autoregressive process of order 1, the representation

$$\begin{aligned}& X_{1t} - 2X_{2t}\\ = \; & \textstyle\sum_{i=0}^{t-1} (1-\alpha_1)^i (\epsilon_1 - 2\epsilon_2)_{t-i} + (1-\alpha_1)^t (X_{10} - 2X_{20})\\ = \; & \textstyle\sum_{i=0}^{\infty} (1-\alpha_1)^i (\epsilon_1 - 2\epsilon_2)_{t-i}\end{aligned}$$

if the initial value is given its invariant distribution as expressed by the past values of the ϵ's. It follows that

$$\begin{aligned}X_{1t} &= 2\textstyle\sum_{i=1}^{t} \epsilon_{2i} + \sum_{i=0}^{\infty} (1-\alpha_1)^i (\epsilon_1 - 2\epsilon_2)_{t-i} + 2X_{20},\\ X_{2t} &= \textstyle\sum_{i=1}^{t} \epsilon_{2i} + X_{20}.\end{aligned}$$

Thus X_t is a nonstationary $I(1)$ variable, and $X_{1t} - 2X_{2t}$ is stationary and $I(0)$, if the initial value $X_{10} - 2X_{20}$ is given its invariant distribution. Hence X_t is cointegrated with the cointegrating vector $(1, -2)$ and

$\sum_{i=1}^{t} \epsilon_{2i}$ is the common trend.

The conclusion of this is that the simple error correction model can generate processes that are nonstationary but cointegrated.

Example 2 shows how the presence of common trends in the moving average representation of X_t can generate cointegration. Example 3 shows that suitable restrictions on the parameters of the autoregressive process will produce cointegration. A general result about the relations between the two approaches was proved by Granger (see Engle and Granger, 1987), and is given in section 2.2.

We have given some examples to illustrate the type of processes we are working with, consisting of a random walk plus a stationary process, and the type of equations that can give rise to such processes. An example from economics is given in section 2.3.

The definitions given above raise a number of interesting mathematical, statistical and probabilistic questions, as well as a number of questions concerning the interpretation of cointegration in the various applications.

- *Mathematical*:

— What kind of nonstationary processes are $I(1)$?

— Which models generate cointegrated processes?

- *Interpretational*:

— How does one formulate interesting economic hypotheses in terms of cointegrating relations?

— What is the interpretation of the cointegrating relations and how can error correction models be usefully applied?

- *Statistical*:

— How does one determine the number of cointegrating relations and common trends?

— How does one estimate the cointegrating relations and the common trends?

— How does one test hypotheses concerning the cointegrating rank?

— How does one test economic hypotheses on the cointegrating relations?

- *Probabilistic*:

— What is the (asymptotic) distribution theory for test statistics and estimators, and can we find improved approximations to the exact distributions?

We discuss some of these questions in the following and illustrate with an application to an economic problem in sections 2.3 and 2.6.

2.2 Granger's representation theorem

This section contains a mathematical discussion of properties of autoregressive processes, in particular the error correction model, with respect to the question of integration, cointegration and common trends. The results are given in Theorem 2.2 and illustrated by some examples.

Only autoregressive processes will be considered, since they form a convenient framework for the statistical analysis. These processes are easy to estimate in the Gaussian case and their properties are well understood.

Consider therefore a general vector autoregressive model for the p-dimensional process X_t defined by the equations

$$X_t = \Pi_1 X_{t-1} + \ldots + \Pi_k X_{t-k} + \epsilon_t,\ t = 1, \ldots, T, \tag{2.1}$$

where $\epsilon_t,\ t = 1, \ldots, T$, are independent variables in p dimensions with mean zero and variance matrix Ω. The values $X_0, \ldots, X_{-k+1}$ are fixed.

Example 4 Consider the process generated by

$$\begin{aligned} \Delta X_{1t} &= \epsilon_{1t}, \\ \Delta X_{2t} &= X_{1t-1} + \epsilon_{2t}, \quad t = 1, \ldots, T. \end{aligned}$$

It is seen that

$$X_{1t} = \sum_{i=1}^{t} \epsilon_{1i} + X_{10},\ t = 1, \ldots, T$$

and hence

$$\begin{aligned} X_{2t} &= \textstyle\sum_{i=1}^{t} X_{1i-1} + \sum_{i=1}^{t} \epsilon_{2i} + X_{20}\ \text{VäVøVÿ} \\ &= \textstyle\sum_{i=1}^{t} \sum_{j=1}^{i-1} \epsilon_{1j} + \sum_{i=1}^{t} \epsilon_{2i} + tX_{10} + X_{20}. \end{aligned}$$

Thus X_{2t} and hence X_t are not $I(1)$ processes since even ΔX_t is nonstationary.

This example shows that even simple autoregressive processes with lag 1 can generate a process which needs two differences to become stationary. Thus we need a theorem that gives precise conditions for an autoregressive process to be an $I(1)$ process.

As usual the properties of the matrix polynomial

$$\Pi(z) = I - \Pi_1 z - \ldots - \Pi_k z^k,$$

determine the properties of the process. We let $|\Pi(z)|$ denote the determinant of $\Pi(z)$, and define the matrices $\Pi = -\Pi(1)$, and $\Gamma =$

$-d\Pi(z)/dz|_{z=1} + \Pi$. We make the following assumption throughout:

Assumption 1 The roots of

$$|\Pi(z)| = 0$$

are either outside the unit disc or at $z = 1$.

Our first result is the classical condition for a process to be stationary; see Anderson (1971, Theorem 5.2.1, p. 170).

Theorem 1 If X_t is given by (2.1) and if Assumption 2.2 holds, then X_t can be given an initial distribution such that it becomes $I(0)$ if and only if Π has full rank, that is, $|\Pi(z)|$ has no unit roots. In this case X_t has the representation

$$X_t = \sum_{i=0}^{\infty} C_i \epsilon_{t-i},$$

where the coefficients are given by $C(z) = \sum_{i=0}^{\infty} C_i z^i = \Pi(z)^{-1}$, $|z| < 1 + \delta$ for some $\delta > 0$.

This result shows that if $|\Pi(z)|$ has all roots outside the unit disc then the process generated by (2.1) is stationary or rather can be made stationary by a suitable choice of the initial distribution. Thus we have to allow other roots of $|\Pi(z)|$ for X_t to be nonstationary.

If unit roots are allowed we can prove another representation, called **Granger's representation theorem**. In order to explain the result, we need, for any $p \times r$ matrix α of rank $r < p$, the notation $\alpha_\perp$ as a $p \times (p - r)$ matrix of full rank such that $\alpha'\alpha_\perp = 0$. We can then formulate the following theorem.

Theorem 2 If X_t is given by (2.1) and if Assumption 1 holds, then X_t is $I(1)$ if and only if

$$\Pi = \alpha\beta', \tag{2.2}$$

where $p \times r$ matrices α, β are of full rank $r < p$, and

$$\alpha_\perp' \Gamma \beta_\perp \text{ has full rank.} \tag{2.3}$$

In this case ΔX_t and $\beta' X_t$ can be given initial distributions such that they become $I(0)$.

The process X_t has the representation:

$$X_t = \beta_\perp \left(\alpha_\perp' \Gamma \beta_\perp\right)^{-1} \alpha_\perp' \sum_{i=1}^{t} \epsilon_i + C(L)\,\epsilon_t + P_{\beta_\perp} X_0,\ t = 1, \ldots, T, \tag{2.4}$$

and satisfies the reduced-form error correction equation

$$\Delta X_t = \alpha\beta' X_{t-1} + \sum_{i=1}^{k-1} \Gamma_i \Delta X_{t-i} + \epsilon_t, \quad t = 1, \ldots, T. \tag{2.5}$$

Thus the cointegrating vectors are β and the common trends are $\alpha'_\perp \sum_{i=1}^{t} \epsilon_i$.

Example 3 (continued) We find in this case

$$\Pi(z) = \begin{pmatrix} \alpha_1 & -2\alpha_1 \\ 0 & 0 \end{pmatrix} + (1-z)\begin{pmatrix} 1-\alpha_1 & 2\alpha_1 \\ 0 & 1 \end{pmatrix},$$

so that

$$\Pi = \begin{pmatrix} -\alpha_1 & 2\alpha_1 \\ 0 & 0 \end{pmatrix} = \begin{pmatrix} -\alpha_1 \\ 0 \end{pmatrix} \begin{pmatrix} 1, & -2 \end{pmatrix}, \; \Gamma = \begin{pmatrix} 1 & 0 \\ 0 & 1 \end{pmatrix},$$

and

$$|\Pi(z)| = (1-z)(1+z\alpha_1 - z).$$

The roots are $z = 1$ and $z = (1-\alpha_1)^{-1}$ if $\alpha_1 \neq 1$. Thus Assumption 1 is satisfied for $0 < \alpha_1 < 2$. We see that Π has reduced rank and find $\alpha' = (-\alpha_1, 0)$ and $\beta' = (1, -2)$. It is seen that $\alpha'_\perp = (0, 1)$ and $\beta'_\perp = (2, 1)$ so that $\alpha'_\perp \Gamma \beta_\perp = 1$, and hence condition (2.3) is satisfied and the process is $I(1)$.

Example 4 (continued) For this example we find

$$\Pi(z) = \begin{pmatrix} 1-z & 0 \\ -z & 1-z \end{pmatrix}, \; |\Pi(z)| = (1-z)^2,$$

which gives the coefficients

$$\Pi = \begin{pmatrix} 0 & 0 \\ 1 & 0 \end{pmatrix}, \Gamma = \begin{pmatrix} 1 & 0 \\ 0 & 1 \end{pmatrix}.$$

Thus $z = 1$ is a double root, Assumption 1 is satisfied and

$$\alpha = \begin{pmatrix} 0 \\ 1 \end{pmatrix}, \; \beta = \begin{pmatrix} 1 \\ 0 \end{pmatrix}, \alpha_\perp = \begin{pmatrix} 1 \\ 0 \end{pmatrix}, \; \beta_\perp = \begin{pmatrix} 0 \\ 1 \end{pmatrix},$$

with the result that

$$\alpha'_\perp \Gamma \beta_\perp = 0.$$

Hence for this example condition (2.3) breaks down, with the result that the process X_t is not $I(1)$ but instead $I(2)$; that is, X_t needs to be differenced twice to become stationary.

For a univariate series the reduced rank condition (2.2) implies that the coefficients of the lagged levels sum to 1, implying that the characteristic polynomial has a unit root, and condition (2.3) then says that the derivative of the polynomial is different from zero at $z = 1$ so that the process has only one unit root. This condition is needed to make sure that the process has to be differenced only once to become stationary.

The parametric restrictions (2.2), (2.3), and Assumption 2.2, determine exactly when the autoregressive equations define an $I(1)$ process that allows for cointegrating relations, and (2.4) gives the representation in terms of common trends or as a random walk plus a stationary process. The equations can be written in error correction form (2.5) and Granger's representation then shows that the common trends formulation (2.4) and the error correction formulation (2.5) are equivalent. Condition (2.3) guarantees that the number of roots of $|\Pi(z)| = 0$ at $z = 1$ equals the rank deficiency of Π, i.e. $p - r$. The proof of Theorem 2.2 is given in Johansen (1991).

We now define a parametric statistical model, the error correction model, given by the vector autoregressive model (2.1) with the restriction (2.2), rewritten as (2.5), which we want to use for describing the statistical variation of the data. The parameters are $(\alpha, \beta, \Gamma_1, \ldots, \Gamma_{k-1}, \Omega)$ which vary freely, giving a total of

$$2pr - r^2 + (k-1)p^2 + \frac{1}{2}p(p+1)$$

parameters. We thus express in parametric form the hypothesis of cointegration, and hypotheses of interest can be formulated as parametric restrictions on the cointegrating relations. What remains is, in connection with applications, to see which hypotheses could be of interest and then to analyse the model in order to find estimators and test statistics and describe their (asymptotic) distributions, but first we turn to a simple economic example to illustrate the ideas.

2.3 Purchasing power parity: an illustrative example

The **law of one price** states that if the same quantity of the same commodity is purchased in two different countries at prices P_1 and P_2 respectively then

$$P_1 = P_2 E_{12}, \tag{2.6}$$

where E_{12} is the exchange rate between the two currencies. It is by no means clear that the same relation will be found if two price indices are compared with the exchange rate. This is due to the definition of the price index, which could vary between countries, and different patterns of consumption in different countries. Still, if this relation is not satisfied approximately there will be pressures on the economy to change either the price levels or the exchange rate. Hence one would expect an adjustment behaviour as modelled by the error correction model.

It is therefore of interest to see in what sense such a relation holds, that is, to look for the so-called **purchasing power parity** (PPP).

The data we analyse was kindly supplied by Ken Wallis and analysed in Johansen and Juselius (1992). It consists of quarterly observations from the first quarter of 1972 to the third quarter of 1987 for the UK wholesale price index, p_1, compared to a trade-weighted foreign price index, p_2, and the UK effective exchange rate, e_{12}. These variables are measured in logs. Also included in the analysis are the three-month Treasury bill rate in the UK, i_1, and the three-month Eurodollar interest rate, i_2. The reason for including the interest rate variables is that one would expect the difference in interest rates to reflect economic expectations concerning changes in exchange rates, that is,

$$i_{1t} - i_{2t} = \Delta^e e_{12.t+1}, \tag{2.7}$$

if there are no restrictions on the movement of capital between countries. Here we have used the symbol Δ^e to indicate economic expectations. If these expectations are model-based (or rational) we replace Δ^e by the conditional expectation of $e_{12.t+1}$ given past information in the data.

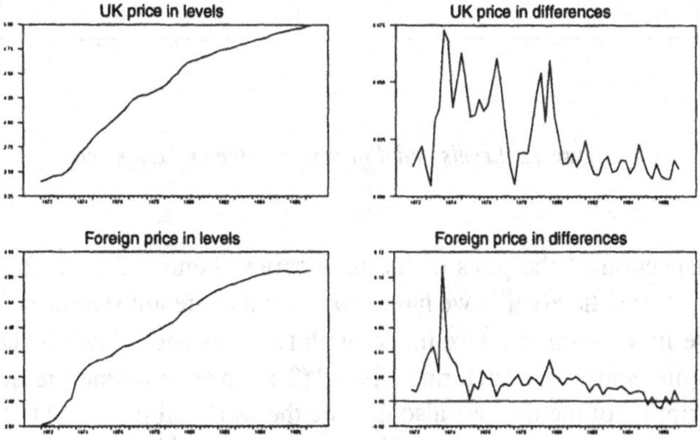

Figure 2.1 *Levels and differences of the prices.*

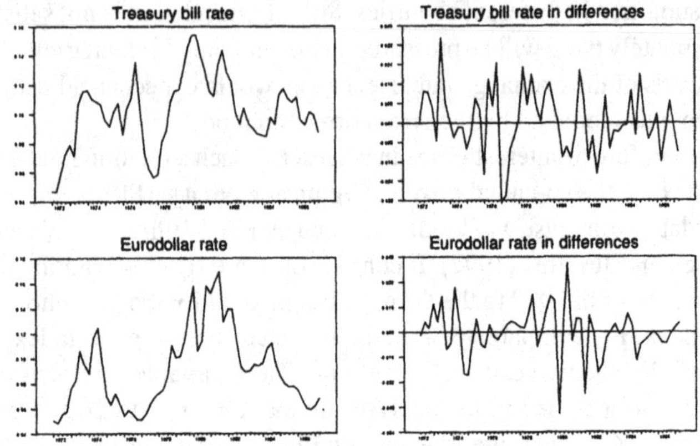

Figure 2.2 *Levels and differences of the interest rates.*

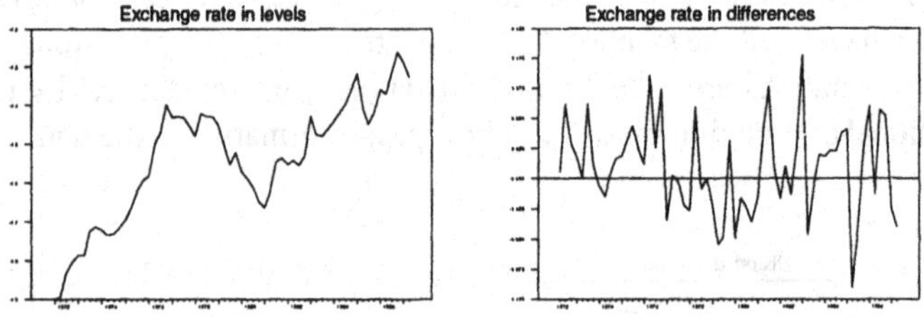

Figure 2.3 *Levels and differences of the exchange rate.*

Inspection of the plots of the time series (Figures 2.1, 2.2 and 2.3) shows immediately that we have processes that are not stationary.

We fit an autoregressive model with two lags and allow for seasonal dummies and a constant term in model (2.1). In order to find a reasonable description of the data we also include the world oil price and treat it as given for the present analysis. This gives rise to added complications in the analysis; we shall not go into these in this paper but refer to Johansen and Juselius (1992) for a full discussion of the application. The residuals (Figures 2.4 and 2.5) show no systematic deviation from independent Gaussian variables and we continue with the analysis of model (2.1).

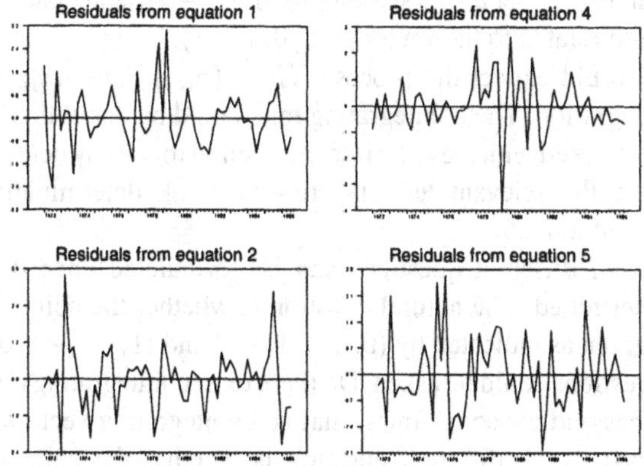

Figure 2.4 *The residuals from the equations for prices and interest rates.*

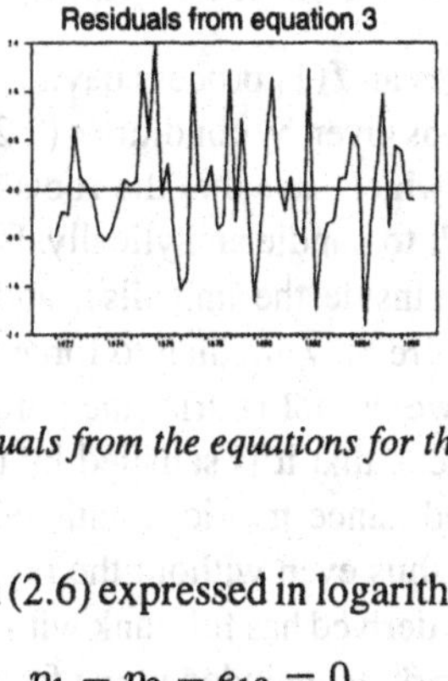

Figure 2.5 *The residuals from the equations for the exchange rate.*

The economic relation (2.6) expressed in logarithms is the PPP relation

$$p_1 - p_2 - e_{12} = 0. \qquad (2.8)$$

This equation is clearly not satisfied by the data. Granger's formulation is that in general a linear combination of variables will be nonstationary, just like individual variables, but one would like to find the most 'stable' or stationary ones and identify them as the interesting economic relations like (2.8). The error correction idea is that changes in prices and exchange rates should be influenced by disequilibrium errors like $(p_1 - p_2 - e_{12})_{t-1}$ through adjustment coefficients. In this formulation the PPP relation (2.8) is translated into a question about the stationarity of the combination $p_{1t} - p_{2t} - e_{12t}$ or, in other words, the question of whether the vector

$(1, -1, -1, 0, 0)$ is a cointegrating vector. Similarly, relation (2.7) is formulated in the model as the stationarity of $i_{1t} - i_{2t}$ and hence turned into a question related to the vector $(0, 0, 0, 1, -1)$.

Thus we would expect the process $X_t' = (p_{1t}, p_{2t}, e_{12t}, i_{1t}, i_{2t})$ to exhibit cointegration. The cointegrating rank could be 2, but this should of course be checked with the data; it seems natural to start by determining the rank, and the relevant test statistics for rank determination are discussed in section 2.5.

A number of different questions can be formulated once the rank has been determined. The natural question is whether the cointegrating vectors are given as indicated by $(0, 0, 0, 1, -1)$ and $(1, -1, -1, 0, 0)$. If not, other questions could be asked. Do there exist cointegrating relations between prices and exchange rates, that is, cointegrating vectors of the form $(a, b, c, 0, 0)$? A different question is whether the cointegrating relations exhibit price homogeneity, that is, if the cointegrating relations have the form $(a, -a, b, c, d,)$.

2.4 Formulation of the reduced-form error correction model and various hypotheses on the cointegrating relations

If model (2.1) describes an $I(1)$ process having cointegration we should restrict the parameters as given by conditions (2.2), (2.3), and Assumption 2.2. Assumption 2.2, which says that the roots are outside the unit disc or at 1, is very difficult to handle analytically. Fortunately it rarely turns out that the roots are inside the unit disc, and if they are, it is more important to know where they are than to force them to the boundary of the unit disc. Hence we do not restrict the parameters in the model by Assumption 1, but check that it is satisfied by the estimates. Condition (2.3) is easily satisfied, since matrices with full rank are dense in the space of all matrices, thus even without the restriction that $\alpha'_\perp \Gamma \beta_\perp$ has full rank the estimator derived has full rank with probability 1. Thus only condition (2.2), $\Pi = \alpha\beta'$, is included in the formulation of the model.

Definition 4 The reduced-form error correction model H_r is described by the equations

$$\Delta X_t = \alpha\beta' X_{t-1} + \sum_{i=1}^{k-1} \Gamma_i \Delta X_{t-1} + \Phi D_t + \epsilon_t, \; t = 1, \ldots, T \quad (2.9)$$

where α and β are $p \times r$, and $\epsilon_1, \ldots, \epsilon_T$ are independent Gaussian $N_p(0, \Omega)$, and the variables D_t are deterministic terms. The parameters $(\alpha, \beta, \Gamma_1, \ldots, \Gamma_{k-1}, \Phi, \Omega)$ are freely varying.

Note that in model H_r the parameters α and β are not identified, since $\Pi = \alpha\beta' = \alpha\xi^{-1}(\beta\xi')'$ for any $r \times r$ matrix ξ of full rank, but that one can estimate the spaces spanned by α and β respectively.

Thus cointegration analysis is formulated as the problem of making inference on the cointegration space, sp(β), and the adjustment space, sp(α). If we want to estimate individual coefficients it is necessary to normalize β or impose restrictions so that the parameters become identified.

Note that we add the condition that the errors are Gaussian in order to work with the likelihood function. We also allow constant terms, linear terms, seasonal dummies, etc., in this formulation of the model. For the asymptotics we have to be a bit more careful (section 2.7). The processes generated by (2.9) contain deterministic terms. The definitions of $I(0)$ and $I(1)$ are concerned only with the stochastic part of the process. In general we define X_t to be $I(0)$ if $X_t - E(X_t)$ satisfies Definition 1. Thus, for instance, a trend-stationary process with a non-zero C-matrix is called an $I(0)$ process. Similarly, the $I(1)$ process X_t is called cointegrating even if $\beta' X_t$ can only be made trend-stationary by a suitable choice of $\beta' X_0$.

The condition that $\Pi = \alpha\beta'$ is sometimes referred to as that of imposing $(p-r)$ unit roots, but we shall think of it as a parametric representation of the existence of (at most) r cointegrating relations. Thus the model is a sub-model of the general autoregressive model defined by the reduced rank hypothesis on the coefficient matrix Π of the levels.

The above allows one to formulate a nested sequence of hypotheses

$$H_0 \subset \ldots \subset H_r \subset \ldots \subset H_p,$$

and the test of H_r in H_p, (see (2.17) below), is then the test that there are (at most) r cointegrating relations. Thus H_0 is just a vector autoregressive model for X_t in differences and H_p the unrestricted autoregressive model for X_t in levels, and the models in between, $H_1, \ldots, H_{p-1}$, give the possibility of exploiting the information in the reduced rank matrix Π. A standard way of analysing nonstationary processes is to difference them to obtain stationarity and then analyse the differences by an autoregressive model. Note that this model is just H_0, the adequacy of which can be tested if we start with the general model H_p.

Once the cointegrating rank has been determined we can test hypotheses about the coefficients α and β, and we next give examples of such hypotheses. It is of considerable importance in work with these models to find a large set of models which can be handled analytically and which seem to give a reasonable description of data. We formulate below a number of such models in connection with the illustrative example that all allow the same analysis. All models are formulated as restrictions on the parameters of model H_r.

The hypothesis that only relative prices enter the cointegrating relations can be expressed as the hypothesis that the coefficients p_1 and p_2 sum to zero, or as the restriction $(1,1,0,0,0)\,\beta = 0$. This is the same restriction on all cointegrating relations which can also be expressed as a direct parametrization

$$\beta = H\varphi, \tag{2.10}$$

where $H = (1,1,0,0,0)'_{\perp}$ is known and $\varphi\,(4 \times r)$ is unknown. This hypothesis on β does not depend on β being identified uniquely, since it is the same set of restrictions on all the relations. If β satisfies (2.10) then so does $\beta\xi$ for any matrix $\xi\,(r \times r)$. Hence (2.10) is a testable hypothesis on the cointegrating space which can be formulated as

$$\mathrm{sp}(\beta) \subset \mathrm{sp}(H).$$

The hypothesis that some cointegrating vectors are known, like $(1,-1,-1,0,0)$ and $(0,0,0,1,-1)$, can be formulated as

$$\beta = (b, \psi)\,, \tag{2.11}$$

where $b\,(p \times r_1)$ is known and $\psi\,(p \times r_2)$ is unknown, $r_1 + r_2 = r$. In particular, it means that the test that an individual variable is stationary can be expressed in the form (2.11) for b equal to a unit vector. Thus the stationarity of a single component of X_t is a special case of cointegration. Note that (2.11) can be formulated as

$$\mathrm{sp}(b) \subset \mathrm{sp}(\beta).$$

A more general linear hypothesis can, for $r = 2$ say, be formulated as

$$\beta = (H_1\varphi_1, H_2\varphi_2)\,, \tag{2.12}$$

where $H_i\ (p \times s_i)$ are known and $\varphi_i\ (s_i \times r_i)$ are unknown and $r_1 + r_2 = r$ (Johansen and Juselius, 1994).

Example 5 An example of (2.12) is given by the hypothesis that p_1, p_2 and e_{12} cointegrate and that the interest rates cointegrate (section 2.3). In this case we are looking for two relations of the form $(a, b, c, 0, 0)$ and $(0, 0, 0, d, e)$, which clearly form a set of uniquely identified equations.

The hypothesis has the form (2.12) with

$$H_1 = \begin{pmatrix} 1 & 0 & 0 \\ 0 & 1 & 0 \\ 0 & 0 & 1 \\ 0 & 0 & 0 \\ 0 & 0 & 0 \end{pmatrix}, \quad H_2 = \begin{pmatrix} 0 & 0 \\ 0 & 0 \\ 0 & 0 \\ 1 & 0 \\ 0 & 1 \end{pmatrix}.$$

Thus we are, in econometric terms, testing for the over-identifying restrictions that there is a cointegrating relation between the variables that has two zeros as coefficients for the interest rates and another one with zeros as coefficients for the prices and exchange rates. It is a common econometric formulation that one wants to identify linear relations of econometric relevance by linear restrictions on the coefficients, in particular zero restrictions.

Consider, for instance, a system consisting of quantity (q), price (p), production costs (c), and income (y). The demand for a good depends on the price and the income, and the supply will depend on the price and the production costs. Thus a demand curve is identified as a relation between the four variables which has a zero coefficient for production costs, and a supply curve is characterized among all possible relations as the one that excludes income.

Thus linear restrictions are formulated on individual relations in the hope that they are sufficiently distinct that identification is in fact possible.

The classical rank condition for identification of a single equation, due to Wald, depends on the parameters of the remaining equations (see Fisher, 1966). One can, however, check that the coefficients in a system with two cointegrating relations, such as (2.12) with $H_i = R_{i\perp}$, $i = 1, 2$, are identified in general by restrictions R_1 and R_2 if $\text{rank}(R_1' H_2) \geq 1$ and $\text{rank}(R_2' H_1) \geq 1$ (see Johansen, 1995a).

Economic insight is used in formulating the problem of interest, and therefore in the choice of variables, as well as in the discussion of which economic relations we expect to find. The statistical model is then used as a description of the nonstationary statistical variation of the data. The cointegrating relations are used as a tool for discussing the existence of long-run economic relations and the various hypotheses are then tested in view of the statistical variation of the data. The interpretation of the cointegrating relations requires a thorough understanding of the underlying economic problem.

2.5 Estimation of cointegrating relations and calculation of test statistics

This section contains a brief description of the regression solution to the problem of cointegrating relations and then discusses how the estimation problem of the various hypotheses from section 2.4 can be solved by analysing the Gaussian likelihood function.

A time-honoured procedure for finding linear relations between two variables Y_t and X_t is to regress Y_t on X_t and then to discuss the properties of the estimator, $\hat{\beta}_{ols}$, under various assumptions on the processes. This was of course the first to be used by Engle and Granger (1987) in their fundamental paper. The problem with the analysis is that since the regressor X_t is generally a nonstationary process the usual simple asymptotic normality does not hold for the estimator.

Stock (1987) proved the at first sight rather surprising result that one gets a superconsistent estimator in the sense that

$$T^{1-\delta}\left(\hat{\beta}_{ols}-\beta\right) \xrightarrow{P} 0, \quad \delta > 0,$$

under the assumption that the regressor is an $I(1)$ process, and that (Y_t, X_t) cointegrate with cointegrating vector $(1, -\beta')$; that is, $Y_t - \beta' X_t$ is stationary. Behind this result is the following very simple idea. In the regression model

$$Y_t = \beta' X_t + \epsilon_t,$$

where ϵ_t are independent Gaussian variables with mean zero and variance σ^2 and the X's are deterministic one finds that $\hat{\beta}_{ols}$ is Gaussian with mean β and variance $\sigma^2(\sum_{t=1}^T X_t X_t')^{-1}$. If the X's are bounded away from zero and infinity then the sum will increase as T and the usual asymptotics hold in the sense that

$$T^{\frac{1}{2}}(\hat{\beta} - \beta)$$

is asymptotically Gaussian. If X_t grows as t, say, then $\sum_{t=1}^T X_t X_t'$ grows as T^3 and $T^{\frac{3}{2}}(\hat{\beta} - \beta)$ is asymptotically Gaussian.

If the ϵ_t's are i.i.d. with mean zero and finite variance, then

$$T^{-\frac{1}{2}} \sum_{i=1}^{[Tu]} \epsilon_i \xrightarrow{w} W(u),$$

where $W(u)$ is Brownian motion. An $I(1)$ process X_t behaves asymptotically like a random walk (see Theorem 2.2), and hence

$$T^{-\frac{1}{2}} X_{[Tu]} \xrightarrow{w} CW(u),$$

with

$$C = \beta_\perp(\alpha'_\perp \Gamma \beta_\perp)^{-1}\alpha'_\perp.$$

By the continuous mapping theorem we find that the proper normalization of the sum of squares is

$$T^{-2}\sum_{t=1}^{T} X_t X_t' = T^{-1}\sum_{t=1}^{T}(T^{-\frac{1}{2}}X_t)(T^{-\frac{1}{2}}X_t)' \xrightarrow{w} \int_0^1 W(u)\,W(u)'du. \tag{2.13}$$

The order (T^2) of the product moments implies that one gets superconsistency of the regression estimator, and the random limit implies that the limiting distribution is not Gaussian, but a rather complicated mixed Gaussian distribution involving Brownian motion and nuisance parameters (section 2.7). Inference for the remaining parameters $\vartheta = (\Gamma_1, \ldots, \Gamma_{k-1}, \alpha, \Phi, \Omega)$ is relatively simple since superconsistency of the estimator for β implies that inference on ϑ can be conducted as if β were known and equal to $\hat{\beta}$, in which case model (2.9) only involves the stationary observables $\hat{\beta}' X_{t-1}$ and the differences of X_t. Section 2.7 gives more details.

This type of result has created a very large literature. For example, see Stock and Watson (1988) for the estimation of the cointegrating rank and the cointegrating relations; Chan and Wei (1988) for inference in unstable processes, and the work of Phillips (1987; 1991) and his co-workers on how to do regression with integrated regressors (Phillips and Durlauf, 1986; Park and Phillips, 1988; 1989; Phillips and Ouliaris, 1990).

The result has lead to a new class of limit distributions, which are combinations of mixed Gaussian and the so-called **unit root distributions**. This type of problem has also been taken up by Jeganathan (1995).

A variant of the regression procedure was suggested by Stock and Watson (1988), who suggested using principal component analysis of the matrix $\sum_{t=1}^{T} X_t X_t'$ to find the linear combinations of the process with the smallest variation as candidates for the cointegrating relations. Box and Tiao (1977) suggested using canonical correlation analysis of X_t with respect to X_{t-1} to pick out the linear combinations that are most easily predicted from the past.

It turns out, however, that the limit distribution for the regression estimator, as well as the estimators involving principal components and canonical correlations of the levels, is very complicated and this makes inference and hypothesis testing difficult. There are ways of eliminating the nuisance parameters by modifying the regression method (Phillips and Hansen, 1990; Park, 1992). Another way of modifying ordinary least squares is to analyse the Gaussian likelihood function and use that as a tool for generating estimators under the various hypotheses investigated in section 2.4. One would expect that if any estimator had a simple limit distribution it would have to be the maximum likelihood estimator. Similarly, one would expect the likelihood ratio test statistic to have a simple limit distribution, even if it turns out only sometimes that we get the χ^2 distribution. Therefore section 2.4 contains a fairly detailed discussion of the model and the hypotheses, and we now turn to inference based on the Gaussian likelihood for the cointegration model.

Model (2.9) gives rise to a reduced rank regression and the solution is available as an eigenvalue problem. It was solved by Anderson (1951) in the regression context and runs as follows.

First we eliminate the parameters $\Gamma_1, \ldots, \Gamma_{k-1}, \Phi$ by regressing ΔX_t and X_{t-1} on $\Delta X_{t-1}, \ldots, \Delta X_{t-k+1}, D_t$. The residuals are R_{0t} and R_{1t} respectively. Next form the sums of squares and products

$$S_{ij} = T^{-1} \sum_{t=1}^{T} R_{it} R_{jt}', \quad i, j = 0, 1.$$

Then the likelihood function maximized with respect to the parameters $\Gamma_1, \ldots, \Gamma_{k-1}, \Phi$ and Ω is given by

$$L_{\max}^{-2/T}(\beta) = (2\pi e)^p |S_{00}| \frac{|\beta'(S_{11} - S_{10} S_{00}^{-1} S_{01})\beta|}{|\beta' S_{11} \beta|}.$$

This is minimized with respect to β by solving the eigenvalue problem

$$|\lambda S_{11} - S_{10} S_{00}^{-1} S_{01}| = 0. \tag{2.14}$$

The solution of this equation gives eigenvalues $1 > \hat{\lambda}_1 > \ldots > \hat{\lambda}_p > 0$ and eigenvectors $\hat{V} = (\hat{v}_1, \ldots, \hat{v}_p)$, which satisfy

$$\lambda_i S_{11} v_i = S_{10} S_{00}^{-1} S_{01} v_i, \quad i = 1, \ldots, p$$

and $\hat{V}' S_{11} \hat{V} = I$.

A maximum likelihood estimator for β is then given by

$$\hat{\beta} = (\hat{v}_1, \ldots, \hat{v}_r). \tag{2.15}$$

An estimator for α is then

$$\hat{\alpha} = S_{01} \hat{\beta},$$

and the maximized likelihood function is given by

$$L_{\max}^{-2/T} = (2\pi e)^p |S_{00}| \prod_{i=1}^{r} \left(1 - \hat{\lambda}_i\right); \tag{2.16}$$

see Johansen (1988) and Johansen and Juselius (1990) for details and applications. One can interpret $\hat{\lambda}_i$ as a squared canonical correlation between ΔX_t and X_{t-1} conditional on $\Delta X_{t-1},\ldots,\Delta X_{t-k+1}$. Thus the estimates of the 'most stable' relations between the levels are those that correlate most with the stationary process ΔX_t corrected for lagged differences and deterministic terms.

Since only $\mathrm{sp}(\beta)$ is identifiable without further restrictions, one really estimates the cointegration space as the space spanned by the first r eigenvectors. This is seen by the fact that if $\hat{\beta}$ is given by (2.15) then $\hat{\beta}\xi$ also maximizes the likelihood function for any choice of ξ $(r \times r)$ of full rank.

This solution provides the answer to the estimation of the models H_r, $r = 0,\ldots,p$. By comparing the likelihoods (2.16) one can test H_r in H_p, i.e. test for r cointegration relations, by the likelihood ratio statistic

$$-2\ln Q\left(H_r|H_p\right) = -T\sum_{i=r+1}^{p}\ln\left(1-\hat{\lambda}_i\right). \tag{2.17}$$

The estimator (2.15) is an estimator of all cointegrating relations and it is sometimes convenient to normalize (or identify) the vectors by choosing a specific coordinate system in which to express the variables in order to facilitate the interpretation and in order to be able to give an estimate of the variability of the coefficients. If c is any $p \times r$ matrix, such that $\beta'c$ has full rank, one can normalize β as

$$\beta_c = \beta\left(c'\beta\right)^{-1},$$

which satisfies $c'\beta_c = I$ provided that $|c'\beta| \neq 0$. A particular example is given by $c' = (I,0)$ and $\beta' = (\beta_1,\beta_2)$ in which case $\beta'c = \beta_1$ and $\beta_c' = \left(I,\beta_1^{-1}\beta_2\right)$ which corresponds to solving the cointegrating relations for the first r variables, if the coefficient matrix of these (β_1) has full rank. The maximum likelihood estimator of β_c is then

$$\hat{\beta}_c = \hat{\beta}\left(c'\hat{\beta}\right)^{-1}.$$

If one wants to estimate β under restrictions this can sometimes be done by the same analysis. Consider the hypothesis (2.10) where $\beta = H\varphi$. In this case

$$\alpha\beta'X_t = \alpha\varphi'H'X_t,$$

which shows that the cointegrating relations are found in $\mathrm{sp}(H)$ by reduced rank regression of ΔX_t on $H'X_{t-1}$ corrected for the lagged differences and D_t, that is, by solving the eigenvalue problem

Table 2.1 *Eigenvectors and eigenvalues for the UK data in section 2.3*

$\hat{\lambda}$	0.40	0.29	0.25	0.10	0.08
p_1	−16.64	−1.68	4.71	9.94	−9.93
p_2	15.12	1.92	−5.99	−23.84	14.42
e_{12}	15.51	5.65	5.24	11.15	4.77
i_1	56.14	−59.17	12.93	−4.06	−22.61
i_2	31.45	55.27	−13.34	29.67	−7.57

$$|\lambda H'S_{11}H - H'S_{10}S_{00}^{-1}S_{01}H| = 0. \tag{2.18}$$

Under hypothesis (2.11) there are some known cointegration relations and in this case $\alpha\beta'X_t = \alpha_1 b'X_t + \alpha_2\varphi'X_t$, which shows that the coefficient α_1 the observable $b'X_{t-1}$ can be eliminated together with the parameters $(\Gamma_1, \ldots, \Gamma_{k-1}, \Phi)$, so that the eigenvalue problem that has to be solved is

$$|\lambda S_{11.b} - S_{10.b}S_{00.b}^{-1}S_{01.b}| = 0, \tag{2.19}$$

where

$$S_{ij.b} = S_{ij} - S_{i1}b\,(b'S_{11}b)^{-1}\,b'S_{1j}, \quad i,j = 0,1.$$

The maximal value of the likelihood function is given by expressions similar to (2.16) and the test of hypotheses (2.10) and (2.11) then consists of comparing the r largest eigenvalues under the various restrictions, since the factor $2\pi e^p|S_{00}|$ cancels.

The hypothesis (2.12) is slightly more complicated, but can be solved by a switching algorithm, where each step involves an eigenvalue problem (see Johansen and Juselius, 1994).

Thus it is seen that a number of interesting hypotheses can be solved provided one has an eigenvalue routine and that one can perform the basic operations on a covariance matrix, namely that of marginalization (transformation) and conditioning. We have programs written in RATS which perform these analyses, and other programs exist written in GAUSS. The packages PC-GIVE and MICROFIT have these analyses implemented.

2.6 The empirical example continued

For the example in section 2.3 we find by solving (2.14) the results in Table 2.1. Note that the first eigenvector has coefficients for p_1, p_2, and

Table 2.2 *Likelihood ratio test statistics* $Q_r = -T\Sigma_{i=r+1}^{5} log(1-\lambda_i)$ *for testing the number r of cointegrating relations*

r	$\hat{\lambda}_{r+1}$	Q_r	95%
4	0.083	5.19	8.08
3	0.102	11.66	17.84
2	0.254	29.26	31.26
1	0.285	49.42	48.41
0	0.401	80.77	69.98

e_{12} which look like the PPP relation, corresponding to relation (2.8), whereas the next two have similar coefficients with opposite sign for the interest rates corresponding to the interest rate differential (2.7).

The challenge in the analysis lies in the interpretation of the eigenvectors corresponding to the largest eigenvalues. One often finds that they have an immediate interpretation, but should remember that what one estimates is the cointegration space spanned by the eigenvectors. Thus one should sometimes 'rotate' or take linear combinations of the vectors to find out what they mean, or still better re-estimate them under (over)identifying restrictions as expressed in (2.12), to check economic hypotheses against data.

We discuss the asymptotic distribution of test statistics and estimators in the next section, but continue here with a brief description of the findings in the data.

The first question is how many cointegrating vectors are consistent with the data. Table 2.2 contains the test statistics defined by (2.17) and their asymptotic 95% quantiles for the hypotheses $H_0, \ldots, H_5$.

It is seen from Table 2.2 that the hypothesis H_0 which gives a test statistic of 80.77 corresponds to an extreme observation in the asymptotic distribution of the test statistic, but that the test statistics for H_1 and H_2 are close to the 95% quantiles. Note that $\hat{\lambda}_2$ and $\hat{\lambda}_3$ are almost identical.

In the further investigation of the data we took $r = 2$, and performed various tests of the type mentioned above. In particular, we found that the hypothesis that both cointegrating vectors have the form $(a, -a, -a, b, c)$ was easily accepted by the data. This is a common restriction, of type (2.10), on both cointegrating relations. When we investigated the hypothesis that the vector $(1, -1, -1, 0, 0)$ was a cointegrating vector it was strongly rejected by the data. Thus our main relation (2.8), the PPP relation, does not hold, but stationarity can be achieved by including the interest rates in the relation. It turns out, however, that the vector $(0, 0, 0, 1, -1)$ is a cointegrating relation, so that the interest rate differential is stationary, in accordance with equation (2.7). Finally, one can ask if p_1, p_2 and e_{12} are cointegrated. This hypothesis is of the form (2.12) and was found not to be supported by the data.

Our understanding of the two-dimensional cointegration space is then the following. One of the vectors is just $(0, 0, 0, 1, -1)$, indicating that the interest rate differential is stationary. If we choose as the other vector the PPP relation $(1, -1, -1, 0, 0,)$ then it is inconsistent with the data, but we have to include the interest rates and can choose a vector of the form $(1, -1, -1, a, b)$, which describes a modified PPP relation. Note that in the system consisting of the relations $(1, -1, -1, a, b)$ and $(0, 0, 0, 1, -1)$ the first is not identified, but if we set either a or b to zero we get a uniquely identified system of cointegrating relations. It turns out (section 2.7) that the likelihood ratio test statistic for such hypotheses is asymptotically χ^2 distributed, so that inference on the long-run relations is easily conducted.

This clearly does not conclude the economic analysis of the data, but the findings are useful for formulating an empirical economic model that describes this small set of variables. Thus the methods presented here are meant only as a tool for investigating long-run relations in the economy.

A structural error correction model has the form

$$\Gamma_0^* \Delta X_t = \alpha^* \beta' X_{t-1} + \sum_{i=1}^{k-1} \Gamma_i^* \Delta X_{t-i} + \Phi^* D_t + \epsilon_t^*,$$

and contains the matrix Γ_0^* which models the short-run simultaneity between the variables. The reduced-form equation is found by dividing through by Γ_0^*. Note that the cointegrating vectors in this model are the same as in the reduced-form equations. This explains why the statistical analysis of cointegrating relations can be performed on the reduced-form equations even though the structural error correction model has not been identified.

Note also that we get an identification problem for β as well as one for the remaining parameters $(\alpha^*, \Gamma_1^*, \ldots, \Gamma_{k-1}^*, \Phi^*)$. They too have to be identified by linear restrictions (or otherwise) in order that the equations can be interpreted.

Thus a cointegrating relation, $\beta' X_t =$ const, can be interpreted as a price relation (if solved for the prices) and describes a long-run equilibrium or steady state, whereas an equation of the structural error correction model describes a price equation (if solved for the changes in the prices) and is interpreted as a model for the dynamic development of prices in interplay with the development of the remaining variables.

It is also of interest to discuss the estimates of the adjustment coefficients α, since they show how the various variables change with past disequilibrium errors, like the modified PPP relation. We found in the present data that changes in foreign prices were not influenced by disequilibrium errors; that is, the α-coefficient was zero for the foreign price equation. This has the consequence that the whole analysis can be performed conditionally on foreign prices, or to put it another way, the foreign prices are weakly exogenous for the cointegrating relations.

Weak exogeneity was introduced by Engle, Hendry and Richard (1983) and is a dynamic version of **S-ancillarity** (Barndorff-Nielsen, 1978) in the following sense. In a time series model for $Z_t = (Y_t', X_t')'$ one wants to give the statistical definition of the notion of exogeneity of X_t for the parameter of interest τ.

Definition 5 The process X_t is called **weakly exogenous** for the parameter of interest τ if there is a parametrization (θ, ξ) of the likelihood function such that

$$L(\theta, \xi) = \prod_{t=1}^{T} g_\theta(Y_t|X_t, Z_1, \dots, Z_{t-1}) \prod_{t=1}^{T} h_\xi(X_t|Z_1, \dots, Z_{t-1}),$$

where $(\theta, \xi) \in A \times B$,and such that $\tau = \tau(\theta)$.

In this case the calculation of the estimator for τ can be based entirely upon the partial likelihood given as the first factor in the likelihood function. In the context of the reduced-form error correction model for Z_t with $\alpha' = (\alpha_y', \alpha_x')$ it is not difficult to see that X_t is weakly exogenous for the parameter of interest β if and only if $\alpha_x = 0$ (see Johansen, 1992).

Thus the fact that foreign prices do not react to disequilibrium errors lends support to their status as an exogenous variable, that is, a variable determined 'outside the system under consideration' which in this case is the UK economy. That $\alpha_x = 0$ implies that the vectors $(0, I)'$ are among the vectors in $\alpha_\perp$. Since these define the common trends, we thus get another interpretation of weak exogeneity, namely that the cumulated unanticipated shocks to the foreign prices constitute one of the common trends.

2.7 Asymptotic theory

This section contains a brief description of the asymptotic theory of test statistics and estimators, as well as a discussion of how the results can be applied to conduct inference about the cointegrating rank and the cointegrating vectors.

The reason why inference for nonstationary processes is interesting and why so many people now work on it, is that it is non-standard, in the sense that estimators are not asymptotically Gaussian and test statistics are not asymptotically χ^2. This was systematically explored by Dickey and Fuller (see Fuller, 1976), in testing for unit roots in univariate processes.

As an example consider the simple model of an autoregressive process of order 1,

$$Y_t = \rho Y_{t-1} + \epsilon_t,$$

where the ϵ_t's are independent Gaussian variables with mean zero and variance σ^2. The null hypothesis of interest is that $\rho = 1$,which implies that Y_t is a random walk, i.e. a nonstationary process. Dickey and Fuller found among other results that when $\rho = 1$, a non-standard limit distribution is obtained, and this can be expressed as

$$T(\hat{\rho} - 1) = T^{-1} \sum_{t=1}^{T} Y_{t-1}\epsilon_t \Big/ T^{-2} \sum_{t=1}^{T} Y_{t-1}^2 \tag{2.20}$$

$$\overset{w}{\to} \int_0^1 W\,dW \Big/ \int_0^1 W^2(u)\,du\,, \tag{2.21}$$

where $W(t)$ is a Brownian motion on [0,1] and the stochastic integral can be calculated as $\int_0^1 W\,dW = \frac{1}{2}\{W(1)^2 - 1\}$. The implication is that the likelihood ratio test statistic is asymptotically distributed as

$$\left(\int_0^1 W\,dW\right)^2 \Big/ \int_0^1 W^2(u)\,du\,.$$

This distribution is often called the 'unit root' or Dickey–Fuller distribution and its multivariate version plays an important role in asymptotic inference for cointegration. We give the main results obtained for likelihood inference, and refer to Johansen (1988; 1991) and Ahn and Reinsel (1990) for the technical details.

Theorem 3 Under the model with $\Phi = 0$ and r cointegrating relations the likelihood ratio statistic (2.17) satisfies

$$\begin{aligned} & -2\ln Q\,(H_r|H_p) \\ = \; & -T\,\Sigma_{i=r+1}^{p} \ln\left(1 - \hat{\lambda}_i\right) \\ \overset{w}{\to} \; & \operatorname{tr}\left\{\int_0^1 B(dB)' \left[\int_0^1 B(u)B(u)'du\right]^{-1} \int_0^1 B(dB)'\right\}. \end{aligned}$$

The process B is a $(p-r)$-dimensional Brownian motion with covariance matrix equal to I. Thus the limit distribution only depends on the number of common trends of the problem. This distribution has then been tabulated by simulation (see Johansen, 1988; Johansen and Juselius, 1990; Reinsel and Ahn, 1990; and Osterwald-Lenum, 1992). It is seen that the distribution is a multivariate generalization of the unit root distribution. This is not surprising, since one can think of testing $\rho = 1$ in the univariate model as a test for no cointegration, i.e. of $r = 0$, when $p = 1$ and $k = 1$.

Although the limit distribution given in Theorem 3 only depends on the degrees of freedom or the dimension of the Brownian motion, it turns out that if a constant term or a linear term is allowed in the model then the limit distribution changes. This leads to a number of complications, as described in Johansen (1994).

It is quite satisfactory, however, for all the other test statistics described in section 2.4 for α and β to have asymptotic χ^2 distributions. Thus the only non-standard test is the test for cointegrating rank. The reason for this is that the asymptotic distribution of the estimator of β is a mixed Gaussian distribution. We give the result for β_c, that is β normalized so that $c'\beta = I$.

Theorem 4 The asymptotic distribution of $\hat{\beta}_c$ is given by

$$T(\hat{\beta}_c - \beta_c) \overset{w}{\rightarrow} (I - \beta_c c')\, \beta_\perp \left\{ \int_0^1 B_1(u) B_1(u)' du \right\}^{-1} \int_0^1 B_1 dB_2', \tag{2.22}$$

where B_1 and B_2 are independent Brownian motions of dimension $p - r$ and r respectively. The asymptotic conditional variance matrix is

$$(I - \beta_c c')\, \beta_\perp \left\{ \int_0^1 B_1(u) B_1(u)' du \right\}^{-1} \times$$
$$\beta_\perp' (I - c\beta_c') \otimes \left(\alpha_c' \Omega^{-1} \alpha_c\right)^{-1}, \tag{2.23}$$

which is consistently estimated by

$$T\left(I - \hat{\beta}_c c'\right) S_{11}^{-1} \left(I - c\hat{\beta}_c'\right) \otimes \left(\hat{\alpha}_c' \hat{\Omega}^{-1} \hat{\alpha}_c'\right)^{-1} \tag{2.24}$$

Thus for a given value of B_1 the limit distribution of $\hat{\beta}$ is just a Gaussian distribution with mean zero and variance given by (2.23). It is this result that implies, by a simple conditioning argument, that the likelihood ratio test statistics for hypotheses about restrictions on β are asymptotically distributed as χ^2 variables, which again makes inference about β very simple if likelihood-based methods are used.

Another way of reading the results (2.22), (2.23) and (2.24) is that since $c'(\hat{\beta}_c - \beta_c) = 0$ we need only consider the coefficients $c_\perp'\left(\hat{\beta}_c - \beta_c\right)$. It now follows from (2.24) that we can act as if these are asymptotically Gaussian with a variance matrix given by

$$Tc_\perp'\left(I - \hat{\beta}_c c'\right) S_{11}^{-1}\left(I - c\hat{\beta}_c'\right) c_\perp \otimes \left(\hat{\alpha}_c'\hat{\Omega}^{-1}\hat{\alpha}_c\right)^{-1},$$

in the sense that this matrix gives the proper normalization of the deviations $T(\hat{\beta}_c - \beta_c)$. Despite the complicated formulation the result is surprisingly simple, since it states only that if β is estimated as identified parameters then the asymptotic variance of $\hat{\beta}$ is given by the inverse information matrix, which is the Hessian used in the numerical maximization of a function. This result is exactly the same as the result that holds for inference in stationary processes. The only difference is the interpretation of (2.24), which for a stationary process would be an estimate of the asymptotic variance, but for $I(1)$ processes is a consistent estimator of the asymptotic conditional variance. The basic property, however, is the same in both cases, namely that it is the approximate scale parameter to use for normalizing the deviation $\hat{\beta} - \beta$.

Table 2.1 contains a lot of information. We use $\hat{\lambda}_{r+1}, \ldots, \hat{\lambda}_p$ to test the model, and $\hat{v}_1, \ldots, \hat{v}_r$ to estimate β. The remaining vectors $\hat{v} = (\hat{v}_{r+1}, \ldots, \hat{v}_p)$, together with $\hat{D} = \text{diag}\left(\hat{\lambda}_1, \ldots, \hat{\lambda}_r\right)$, contain information on the asymptotic conditional variance matrix of the estimate $\hat{\beta}$. This can be exemplified in the form of a Wald test. One can prove that under the hypothesis $K'\beta = 0$, which is of the form (2.10) for $H = K_\perp$,

$$T\text{tr}\left\{K'\hat{\beta}\left(\hat{D}^{-1} - I\right)^{-1}\hat{\beta}'K\left(K'\hat{v}\hat{v}'K\right)^{-1}\right\}$$

is asymptotically χ^2 distributed. Thus, in particular, if $r = 1$ and $\beta = (v_{11}, \ldots, v_{p1})$, and one wants to see if the coefficient v_{11} is significant, one can calculate the quantity

$$T^{\frac{1}{2}}\hat{v}_{11}\left\{\left(\hat{\lambda}_1^{-1} - 1\right)\left(\Sigma_{j=2}^p \hat{v}_{1j}^2\right)\right\}^{-\frac{1}{2}},$$

and compare it to the quantiles of a standard Gaussian distribution (see Johansen, 1991).

One can now discuss why inference about β becomes difficult when based on the simple regression estimator. This is because the limiting distribution of $\hat{\beta}_{ols}$ is expressed as an integral as in Theorem 4, but with dependent B_1 and B_2. This again implies that for given B_1 the limit distribution of the estimator does not have conditional mean zero, which implies that the test statistics based upon the regression estimator will have some non-central distribution with nuisance parameters.

Inference for the remaining parameters $\vartheta = (\alpha, \Gamma_1, \dots, \Gamma_{k-1}, \Omega)$ is different. This is explained by Phillips (1991) and the idea is roughly as follows. The second derivative of the log-likelihood function with respect to β tends to infinity as T^2 (see (2.13)), whereas the second derivative with respect to ϑ and the mixed derivatives tend to infinity as T. This means that $\hat{\beta} - \beta$ has to be normalized by T and $\hat{\vartheta} - \vartheta$ by $T^{\frac{1}{2}}$. This, on the other hand, requires a normalization of the mixed derivatives by $T^{3/2}$ and makes them disappear in the limit. Thus in the limit the information matrix, which is used to normalize $(\hat{\beta} - \beta, \hat{\vartheta} - \vartheta)$, is block diagonal with one block for β and one block for the remaining parameters ϑ. Thus inference concerning β can be conducted as if ϑ were known and vice versa (see Johansen, 1991).

2.8 Conclusion

We have shown that the notion of cointegration as a way of describing long-run economic relations can be formulated in the autoregressive framework as the hypothesis of reduced rank of the coefficient matrix of the levels. This allows explicit maximum likelihood estimation of the cointegrating relations, both unrestricted and under certain types of linear restriction that seem to correspond to interesting economic hypotheses. We have found the asymptotic distribution of the likelihood ratio test statistic for the cointegrating rank and tabulated it by simulation. It is shown that restrictions on β can be tested using the χ^2 distribution. This shows that the property of nonstationarity of the processes, instead of being a nuisance to be eliminated by differencing, can be used as a strong tool to investigate long-run dependencies between variables.

The most important problem for further studies is, in my opinion, the problem of deriving better approximations for the distributions than those given by the asymptotic results. It often turns out that with, say, 100 observations the tables provided give very poor approximations to the actual distributions. Thus there is a need for finding Bartlett corrections for the various test statistics.

Another important area of research is to extend these methods to processes that are better described as integrated of order 2. For these models the asymptotic theory becomes more difficult, although some results have been found (see Johansen, 1995c; Paruolo, 1995).

Acknowledgements

This paper is based on the Cramér Lecture presented at the 2nd World Conference of the Bernoulli Society for Mathematical Statistics and Probability at Uppsala in 1990, and the Neyman lecture presented at the IMS meeting at Boston in 1992. I would like to thank anonymous referees as well as David Cox for careful reading and many suggestions for improving the presentation. Throughout the period of this research I have been partially supported by the Danish Social Sciences Research Council.

References

Ahn, S.K. and Reinsel, G.C. (1990) Estimation for partially nonstationary multivariate autoregressive models. *Journal of the American Statistical Association*, **85**, 813–823.

Anderson, T.W. (1951) Estimating linear restrictions on regression coefficients for multivariate normal distributions. *Annals of Mathematical Statistics*, **22**, 327–351.

Anderson, T.W. (1971) *The Statistical Analysis of Time Series.* Wiley, New York.

Banerjee, A., Dolado, J.J., Galbraith, J.W. and Hendry, D.F. (1993) *Co-integration, Error Correction, and the Econometric Analysis of Non-stationary Data.* Oxford University Press, Oxford.

Barndorff-Nielsen, O.E. (1978) *Information and Exponential Families in Statistical Theory.* Wiley, New York.

Box, G.P.E. and Tiao, G.C. (1977) A canonical analysis of multiple time series. *Biometrika*, **64**, 355–365.

Chan, N.H. and Wei, C.Z. (1988) Limiting distributions of least squares estimates of unstable autoregressive processes. *Annals of Mathematical Statistics*, **16**, 367–410.

Cuthbertson, K., Hall, S.G. and Taylor, M.P. (1992) *Applied Econometric Techniques.* Harvester Wheatsheaf, New York.

Engle, R.F. and Granger, C.W.J. (1987) Co-integration and error correction representation, estimation and testing. *Econometrica*, **55**, 251–276.

Engle R.F. and Granger, C.W.J. (1991) *Long-run Econometric Relations. Readings in Cointegration.* Oxford University Press, Oxford.

Engle, R.F., Hendry, D.F. and Richard, J.-F. (1983) Exogeneity. *Econometrica*, **51**, 277–304.

Fisher, F.M. (1966) *The Identification Problem in Econometrics.* McGraw-Hill, New York.

Fuller W. (1976) *Introduction to Statistical Time Series.* Wiley, New York.

Granger, C.W.J. (1983) Cointegrated variables and error correction models, UCSD Discussion paper 83-13a.

Hamilton, J. (1994) *Time Series Analysis.* Princeton University Press, Princeton, NJ.

Hargreaves, C.P. (ed.) (1994) *Nonstationary Time Series Analysis and Cointegration.* Oxford University Press, Oxford.

Jeganathan, P. (1995) Some aspects of asympotic theory with applications to time series models. To appear in *Econometric Theory.*

Johansen, S. (1988) Statistical analysis of cointegration vectors. *Journal of Economic Dynamics and Control*, **12**, 231–254.

Johansen, S. (1991) Estimation and hypothesis testing of cointegration vectors in Gaussian vector autoregressive models. *Econometrica*, **59**, 1551–1580.

Johansen, S. (1992) Cointegration in partial systems and the efficiency of single equation analysis. *Journal of Econometrics*, **52**, 389–402.

Johansen, S. (1994) The role of the constant and linear terms in cointegration analysis of nonstationary variables. *Econometric Reviews*, **13**, 205–229.

Johansen, S. (1995a) Identifying restrictions of linear equations – with applications to simultaneous equations and cointegration. *Journal of Econometrics*, **69**, 111–132.

Johansen, S. (1995b) *Likelihood Based Inference in Cointegrated Vector Autoregressive Models.* Oxford University Press, Oxford. To appear.

Johansen, S. (1995c) A statistical analysis of cointegration for $I(2)$ variables. *Econometric Theory*, **11**, 25–59.

Johansen, S. and Juselius, K. (1990) Maximum likelihood estimation and inference on cointegration – with applications to the demand for money. *Oxford Bulletin of Economics and Statistics*, **52**, 169–210.

Johansen, S. and Juselius, K. (1992) Structural hypotheses in a multivariate cointegration analysis of the PPP and UIP for UK. *Journal of Econometrics*, **53**, 211–244.

Johansen, S. and Juselius, K. (1994) Identification of the long-run and the short-run structure. An application to the ISLM model. *Journal of Econometrics*, **63**, 7–36.

Lütkepohl, H. (1993) *Introduction to multiple times series analysis.* Springer-Verlag, New York.

Osterwald-Lenum, M. (1992) A note with fractiles of the asymptotic distribution of the maximum likelihood cointegration rank test statistic. Four cases. *Oxford Bulletin of Economics and Statistics* **54**, 461–471.

Park, J.Y. (1992) Canonical cointegrating regressions. *Econometrica*, **60**, 119–143.

Park, J.Y. and Phillips, P.C.B. (1988) Statistical inference in regressions with integrated processes. Part 1. *Econometric Theory,* **4**, 468–498.

Park, J.Y. and Phillips, P.C.B. (1989) Statistical inference in regressions with integrated processes. Part 2. *Econometric Theory,* **5**, 95–131.

Paruolo, P. (1995) Asymptotic efficiency of the 2 step estimator in $I(2)$ VAR systems. To appear in *Econometric Theory.*

Phillips, A. W. (1954) Stabilisation policy in a closed economy. *Economic Journal*, **64**, 290–323.

Phillips, P.C.B. (1987) Time series with a unit root. *Econometrica*, 55, 277–301.

Phillips, P.C.B. (1991) Optimal inference in cointegrated systems. *Econometrica,* **59**, 283–306.

Phillips, P.C.B. and Durlauf, S.N. (1986) Multiple time series regression with integrated processes. *Review of Economic Studies,* **53**, 473–495.

Phillips, P.C.B. and Hansen, B.E. (1990) Statistical inference on instrumental variables regression with $I(1)$ processes. *Review of Economic Studies*, **57**, 99–124.

Phillips, P.C.B. and Ouliaris, S. (1990) Asymptotic properties of residual based tests for cointegration. *Econometrica*, **58**, 165–193.

Reinsel G.C. and Ahn, S.K. (1990) Vector autoregressive models with unit roots and reduced rank structure, estimation, likelihood ratio test, and forecasting. *Journal of Time Series Analysis*, **13**, 283–295.

Sargan, J.D. (1964) Wages and prices in the United Kingdom. A study in econometric methodology. In Hart, P.E., Mills, G. and Whitaker, J.K. (eds), *Econometric Analysis for National Economic Planning*, Butterworths, London.

Stock, J.H. (1987) Asymptotic properties of least squares estimates of cointegration vectors. *Econometrica*, **55**, 1035–1056.

Stock, J.H. and Watson, M.W. (1988) Testing for common trends. *Journal of the American Statistical Association,* **83**, 1097–1107.

CHAPTER 3

Forecasting in macro-economics

Michael P. Clements and David F. Hendry

3.1 Introduction

This Chapter draws upon a number of our earlier publications to highlight the key issues and problems facing a theory of forecasting in macro-economics. In particular, we draw on Clements and Hendry (1993a; 1994a; 1994b; 1995a) and Hendry and Clements (1994a, 1994b). Because social behaviour can change in an economy which is high-dimensional and evolving, and regime shifts occur, mechanically derived forecasts (that is, forecasts made without human intervention) cannot usually successfully predict outcomes. Macro-econometric models seek to represent the main relationships determining economic behaviour in a structured framework, heavily constrained by many inherent identities linking groups of outcomes. Their coefficients are estimated from available historical data, then the models are solved given recent past data and perhaps assumptions about policy and external variables to make sequences of forecasts over a range of future periods. The resulting forecasts usually represent a compromise between the model's output and the intuition and experience of the modeller.

The treatment in Haavelmo (1944) of forecasting in a probabilistic framework was the progenitor of the current textbook approach to forecasting in econometrics. Suppose there are T observable values $\mathbf{X}_T^1 = (\mathbf{x}_1, \ldots, \mathbf{x}_T)$ on a vector random variable, from which to predict the H future values $\mathbf{X}_{T+H}^{T+1} = (\mathbf{x}_{T+1}, \ldots, \mathbf{x}_{T+H})$. Let the joint probability of the observed and future $\mathbf{x}$'s be $D(\mathbf{X}_{T+H}^1 | \mathbf{X}_0, \theta)$, where the form of $D(\cdot)$ is assumed known, $\theta \in \Theta \subseteq \mathcal{R}^p$ is the parameter vector, and $\mathbf{X}_0$ denotes the initial conditions pre-sample. Denote the probability distribution of the future $\mathbf{x}$'s conditional on past $\mathbf{x}$'s by $D_2(\mathbf{X}_{T+H}^{T+1} | \mathbf{X}_T^1, \theta)$, and the probability distribution of the observed $\mathbf{x}$'s by $D_1(\cdot)$. Then, factorizing into conditional and marginal probabilities

$$D\left(\mathbf{X}_{T+H}^{1} \mid \mathbf{X}_0, \theta\right) = D_2\left(\mathbf{X}_{T+H}^{T+1} \mid \mathbf{X}_T^1, \mathbf{X}_0, \theta\right) \times D_1\left(\mathbf{X}_T^1 \mid \mathbf{X}_0, \theta\right). \tag{3.1}$$

Thus, for any realization of the observed $\mathbf{x}$'s, denoted by the set $\mathcal{E}_1 \subset \mathcal{R}^T$, the probability that the future $\mathbf{x}$'s will lie in any $\mathcal{E}_2$, where $\mathcal{E}_2 \subset \mathcal{R}^H$, can be calculated from $D_2(\cdot)$. In practice, $D_2(\cdot)$ is unknown and must be estimated from the realization $\mathcal{E}_1$, which requires the 'basic assumption' that: 'The probability law $D(\cdot)$ of the $T + H$ (vector) variables $(\mathbf{x}_1, \ldots, \mathbf{x}_{T+H})$ is of such a type that the specification of $D_1(\cdot)$ implies the complete specification of $D(\cdot)$ and, therefore, of $D_2(\cdot)$' (Haavelmo, 1994, p.107: our notation).

When the econometric model and the mechanism generating the data coincide in an unchanging world, the theory of economic forecasting is reasonably well developed, and a number of propositions follow. A forecast is a function of past information, and for period $T + h$, conditional on information up to period T, is given by $\hat{\mathbf{x}}_{T+h} = \mathbf{f}_T\left(\mathbf{X}_T^1\right)$ where $\mathbf{f}_T$ may depend on a prior estimate of θ. Forecasts calculated as the conditional expectation given the model are optimal in that they are unbiased and efficient (any other predictor has a larger mean squared forecast error). If the model is the data generation process, then the expected future value of the variables, given all information available at the present time, is the conditional expectation with respect to the model. Forecasts and future realizations of the process will differ only because the process generating the data is stochastic, with an error component in the data generation process which is unpredictable from past information (an innovation): see Granger and Newbold (1977).

Unfortunately, the historical track record is littered with less than brilliant forecasts. There have been major episodes of predictive failure when model forecasts have systematically over- or underpredicted for substantial periods, and realized outcomes have been well outside any reasonable *ex ante* confidence interval computed from the uncertainties due to parameter estimation and lack of fit. Examples include the forecasts made by the major UK model-based forecasting teams over the 1974–5 and 1979–81 recessions, discussed by Wallis (1989).

This suggests that it is inappropriate to use a theory of forecasting based on assuming a stationary process with constant parameters which are accurately captured by the model. Nevertheless, the majority of analyses adopt just such a perspective, and concentrate on a world in which a constant-parameter, unchanging relation such as (3.1) holds. It is often assumed that stationarity may be achieved by differencing the data sufficiently (see, for example,Kendall, Stuart and Ord, 1987), ruling out the possibility of nonstationarities other than integratedness. Further, as argued by Fildes and Makridakis (1994), there is a departure between the implications of the theory and the associated empirical findings.

In terms of (3.1), then, we might want to allow that both $D(\cdot)$ and θ may change between sample (θ_1) and forecast periods (θ_2), so the operational forecast procedure using

$$D_1\left(\mathbf{X}_{T+H}^{T+1} \mid \mathbf{X}_T^1, \mathbf{X}_0, \theta_1\right)$$

may provide a non-optimal basis for decision-takers. The realistic situation is probably even worse in that inconsistent estimates of θ may be available, and $\mathbf{X}_T$ could be inaccurately or provisionally estimated at the time of the forecast.

In this review, we focus on some of the implications of forecasting in a nonstationary and changing world, where the model and mechanism differ. The situation is not as hopeless as this description may make it seem. Surprisingly, a somewhat discredited procedure transpires to be justifiable, namely a class of modifications to forecasts called 'intercept corrections'. In certain respects, therefore, Klein (1971) is the most direct precursor of the approach we adopt.

Section 3.2 describes the framework within which we analyse economic forecasting. Section 3.3 notes some alternative methods of forecasting. The economic system and forecasting models are formulated in section 3.4, and measures of forecast accuracy discussed in section 3.5. Section 3.6 develops a taxonomy of forecast errors. Cases with parameter constancy and non-constancy are investigated in sections 3.7 and 3.8 respectively, followed by a discussion of the potential role of intercept corrections in section 3.9. Section 3.10 concludes.

3.2 A framework for economic forecasting

There are many attributes of the economic system to be forecast that impinge on the choice of method by which to forecast and on the likelihood of obtaining reasonably accurate forecasts. Here we distinguish six different facets: [A] the nature of the data generation process; [B] the knowledge level about that data generation process; [C] the dimensionality of the system under investigation; [D] the form of the analysis; [E] the forecast horizon; and [F] the linearity or otherwise of the system. Then we have:

[A] Nature of the data generation process

[*i*] stationary;
[*ii*] cointegrated stationary;
[*iii*] evolutionary, nonstationary.

[B] Knowledge level

[*i*] known data generation process, known θ;
[*ii*] known data generation process, unknown θ;
[*iii*] unknown data generation process, unknown θ.

[C] Dimensionality of the system

[*i*] scalar process;
[*ii*] closed vector process;
[*iii*] open vector process.

[D] Form of analysis

[*i*] asymptotic analysis;
[*ii*] finite sample results.

[E] Forecast horizon

[*i*] one-step;
[*ii*] multistep.

[F] Linearity of the system

[*i*] linear;
[*ii*] nonlinear.

An exhaustive analysis under this taxonomy would generate 216 cases! Here, we consider only a few states corresponding to the joint selection of elements from each of $\{\mathsf{A}, \mathsf{B}, \mathsf{C}, \mathsf{D}, \mathsf{E}, \mathsf{F}\}$, chosen to allow analyses of forecasting that capture the salient features of the environment in which the forecaster habitually operates. Thus, from $\{\mathsf{A}\}$ we choose $\{iii\}$, in the belief that periods of economic turbulence resulting from structural breaks or regime shifts historically go hand-in-hand with episodes of dramatic predictive failure. For example, the 1974–5 and 1979–81 recessions, noted in section 3.1, were at least in part due to the behaviour of OPEC. Other examples abound. The correlation between turbulence and predictive failure is to be expected, but what is surprising is that most analyses of economic forecasting are firmly rooted in the assumption of a constant data generation process. For an econometric theory of forecasting to deliver relevant conclusions about empirical forecasting, it must be based on assumptions that adequately capture the appropriate aspects of the real world to be forecast.

Under {B} we move between {*i*}, {*ii*} and {*iii*} to make various points but recognize that in practice the data generation process is not known, and that the best that may be achieved is a 'congruent' model (see section 3.4). Thus, the 'textbook' assumption that the model and data generation process are the same is invariably false.

Under {C} we consider closed vector processes, so that all non-deterministic variables are forecast within the system. In open systems not all the stochastic variables are modelled, but nevertheless off-line projections of non-modelled variables have to be made, so that the distinction between {*ii*} and {*iii*} may seem somewhat artificial. The distinction may be useful if the non-modelled variables are primarily pre-announced policy variables over the relevant forecast horizon, so that closing the model by specifying autoregressions for such variables would overstate the uncertainty surrounding their future values. We abstract from such issues here, but Granger and Deutsch (1992) discuss policy variables.

The choice of {*i*} from {F} is mainly for analytical tractability. Salmon and Wallis (1982), Brown and Mariano (1984; 1989), Mariano and Brown (1991), and Granger and Teräsvirta (1993) provide references to the forecasting literature for nonlinear models. However, we believe the implications of our analysis are not greatly altered by nonlinearity. We often use asymptotic analyses {D*i*} and make approximations where necessary to uncover the main factors at work. We generally consider multistep forecasts {E*ii*} since one-step forecast performance is only a reliable guide to multistep performance under certain stringent conditions, which are unlikely to hold in practice, such as that the forecast model coincides with the data generation process (see, for example, Fama and French, 1988; Baillie, 1993).

3.3 Alternative methods of forecasting

There are many ways of making economic forecasts besides using econometric models. The success of all forecasting methods requires that there are regularities to be captured, and that such regularities are informative about the objects to be forecast. The success of a particular method will depend on how well that method captures those regularities and (less obviously) excludes non-regularities. The analysis in section 3.8 illustrates these propositions for model-based forecasting.

Methods of forecasting include guessing, 'rules of thumb' or 'informal models', naive extrapolation, leading indicators, surveys, time series models, and econometric systems. If we insist upon the methods being both open to adversarial scrutiny and replicable, then the first two are ruled out. Naive extrapolation refers to the simple projection of recent trends. Its success requires that the perceived tendencies do indeed persist, corresponding to the informative regularities, and that those which do not persist are excluded. Extrapolative predictors are fine during epochs when the economy happens to evolve in an 'extrapolative fashion', but otherwise are unhelpful since no attempt is made to capture the complex interactions and interrelationships that may generate changed behaviour. Forecasting based on current, leading and lagging indicators is undergoing a surprising revival of interest: see Stock and Watson (1989; 1992) and the collection edited by Lahiri and Moore (1991); Emerson and Hendry (1994) provides a critical view. Survey information may be useful, but typically is only available for short horizons, and even then will be more or less accurate to the extent that agents' plans are fulfilled, an aspect which itself may benefit from formal modelling. Also, it is not clear how to interpret point forecasts in the absence of any guidance as to their accuracy; see Chatfield (1993) on the importance of interval forecasts.

Time-series and econometric methods share the advantage of being based on statistical models, so that formal measures of uncertainty, such as confidence intervals, can be calculated for their forecasts. Harvey (1989) shows that many popular, apparently *ad hoc* forecasting techniques, such as exponentially weighted moving averages and the schemes of Holt (1957) and Winters (1960), can be derived from the class of statistically well-founded 'structural' time series models. Such methods effectively fit linear functions of time to the series but place greater weight on the more recent observations. The intuitive appeal of such schemes stems from the belief that the future is more likely to resemble the recent than the remote past.

Scalar versions of time series models usually take the form proposed by Kalman (1960) or Box and Jenkins (1976). There are good reasons why integrated autoregressive moving average models (ARIMAs) might be regarded as the dominant class of time series models (where the order of the integrated component is d, the minimum number of times the series has to be differenced to be stationary). The Wold decomposition theorem (Wold, 1938) states that any purely indeterministic stationary time series can be expressed as an infinite moving average (MA); see Cox and Miller (1965, pp. 286–288), for a lucid discussion. Moreover, an infinite MA can be approximated to any required degree of accuracy by an ARMA model. Typically the order of the AR and MA polynomials (p and q) required to fit the series adequately may be relatively low. Many economic time series may be nonstationary, but provided they can be made stationary by differencing then they are amenable to analysis within the Box–Jenkins framework. Similarly, at least in principle, deterministic components such as linear functions of time can be readily dealt with by prior regression, so that Box–Jenkins is then applied to the remainder.

Although historically ARIMA models have performed well relative to econometric methods, Harvey (1989, pp. 80–81), notes some of the problems with ARIMA modelling. Part of their success is probably due to dynamic misspecification in econometric models, but such a source of error was greatly reduced during the 1980s as modern methods were adopted; and part derives from the type of model adopted, as discussed below. He also considers time series modelled as the sum of unobserved components which have a direct interpretation as trend, seasonal and irregular components. Typically, such models contain a number of disturbance terms, but from the structural time series models, reduced forms with a single disturbance term can be derived. The reduced forms are Box–Jenkins ARIMA models, although the derivation from a structural model implies certain restrictions on the parameters of the ARIMA model. Harvey (1989, pp. 68–70), obtains the ARIMA models implied by the principal structural models.

The multivariate successor to Box–Jenkins is the vector autoregressive representation (see Doan, Litterman and Sims, 1984). In the USA this approach has claimed some successes.

Formal econometric systems of national economies fulfil many roles other than just being devices for generating forecasts; for example, such models consolidate existing empirical and theoretical knowledge of how economies function, provide a framework for a progressive research strategy, and help explain their own failures. Econometric models and multivariate time series models are taken to be the primary methods of economic forecasting in this paper.

3.4 The economic system and forecasting models

The vector of n variables of interest is denoted by $\mathbf{x}_t$, and its data generation process is taken to be the first-order vector autoregression

$$\mathbf{x}_t = \tau + \boldsymbol{\Upsilon}\mathbf{x}_{t-1} + \nu_t, \tag{3.2}$$

where $\boldsymbol{\Upsilon}$ is an $n \times n$ matrix of coefficients, and τ is an n-dimensional vector of constant terms. The error $\nu_t \sim IN_n(0, \boldsymbol{\Omega})$, with expectation $E[\nu_t] = 0$ and variance matrix $\text{var}[\nu_t] = \boldsymbol{\Omega}$. The first-order vector autoregression in (3.2) can be interpreted as the companion form to a qth-order vector autoregression

$$
\begin{aligned}
\mathbf{x}_t &= \begin{pmatrix} \underline{\mathbf{x}}_t \\ \underline{\mathbf{x}}_{t-1} \\ \vdots \\ \underline{\mathbf{x}}_{t-q+1} \end{pmatrix} = \begin{pmatrix} \underline{\tau} \\ 0 \\ \vdots \\ 0 \end{pmatrix} \\
&+ \begin{pmatrix} \boldsymbol{\Upsilon}_{\{1\}} & \boldsymbol{\Upsilon}_{\{2\}} & \cdots & \boldsymbol{\Upsilon}_{\{q-1\}} & \boldsymbol{\Upsilon}_{\{q\}} \\ \mathbf{I}_n & 0 & \cdots & 0 & 0 \\ 0 & \mathbf{I}_n & 0 & 0 & \vdots \\ \vdots & \ddots & \ddots & \ddots & \vdots \\ 0 & \cdots & 0 & \mathbf{I}_n & 0 \end{pmatrix} \begin{pmatrix} \underline{\mathbf{x}}_{t-1} \\ \underline{\mathbf{x}}_{t-2} \\ \vdots \\ \underline{\mathbf{x}}_{t-q} \end{pmatrix} \\
&+ \begin{pmatrix} \underline{\nu}_t \\ 0 \\ \vdots \\ 0 \end{pmatrix}.
\end{aligned}
$$

Thus, the lag length assumption is not restrictive, although our main aim here is expository. Linearity is more constraining (see Mariano and Brown, 1983), but we doubt if it is the main explanation of recent predictive failures. Conversely, (3.2) allows for integrated and cointegrated processes (see below), and is a useful framework for considering sources of forecast errors.

Although the *form* of the model to be fitted coincides with (3.2), in so far as it is a linear representation linking elements of $\mathbf{x}_t$, its specification could differ in every important regard from that of the data generation process, due to imposing invalid restrictions on the parameters. We write the model as

$$\mathbf{x}_t = \tau_p + \boldsymbol{\Upsilon}_p \mathbf{x}_{t-1} + \mathbf{u}_t, \tag{3.3}$$

where the parameter estimates $(\hat{\tau} : \hat{\boldsymbol{\Upsilon}} : \hat{\boldsymbol{\Omega}})$ are possibly inconsistent, with $\tau_p \neq \tau$ and $\boldsymbol{\Upsilon}_p \neq \boldsymbol{\Upsilon}$, because of the model misspecification. Given knowledge of the data generation process, it is possible to deduce the values of the model parameters and the properties of the model's error term $\mathbf{u}_t$. This is formalized in the theory of reduction, which explains the origin and status of empirical econometric models in terms of the implied information reductions relative to the process that generated the data. Reductions implicitly transform the parameters of the process. Thus, empirical models are reductions of the data generation process, not numerically calibrated theoretical models, with error processes which are derived by reduction operations and are not autonomous; see, among many others, Hendry and Richard (1982; 1983), Gilbert (1986), Spanos (1986), Hendry (1987; 1993; 1995), Ericsson, Campos and Tran (1990), Florens, Mouchart and Rolin (1990), Hendry and Mizon (1990; 1993), Cook and Hendry (1993).

Schematically, if the empirical model is

$$\underset{\text{[observed]}}{y_t} = \underset{\text{[explanation]}}{g(z_t)} + \underset{\text{[remainder]}}{\epsilon_t} \tag{3.4}$$

then the left-hand side determines the right, rather than the other way round. Thus y_t can be decomposed into two components, $g(z_t)$ (a part which can be explained by z) and ϵ_t (a part which is unexplained). Such a partition is possible even when y_t does not depend on $g(z_t)$, but is determined by completely different factors $h(x_t)$, say. In econometrics,

$$\epsilon_t = y_t - g(z_t)$$

describes empirical models: changing the choice or specification of z_t on the right-hand side alters the left-hand side, so $\{\epsilon_t\}$ is a derived process. $\{\epsilon_t\}$ in (3.4) is not a random drawing from nature: it is what is left over from y_t after extracting $g(z_t)$ and must represent everything omitted from the model, including all the mistakes in formulating $g(\cdot)$ or selecting z.

The resulting error process could be autocorrelated and heteroscedastic. However, empirical models are susceptible to design to achieve desired criteria (such as white noise errors). Statistics used in that design process become part of the selection criteria, perhaps as diagnostics used to check on the validity of the reduction, and hence cease to be tests in any useful sense. Thus, it will often be the case in (3.3) that

$$\mathbf{u}_t = \mathbf{x}_t - E_{\mathsf{M}}[\mathbf{x}_t \mid \mathbf{x}_{t-1}]$$

(where $E_{\mathsf{M}}[\cdot|\cdot]$ denotes a conditional expectation with respect to the model) is an innovation process

$$E[\mathbf{u}_t \mid \mathbf{x}_{t-1}] = 0.$$

Two models we shall consider are defined by ($\tau_p = \tau$, $\boldsymbol{\Upsilon}_p = \boldsymbol{\Upsilon}$) and ($\tau_p = \gamma$, $\boldsymbol{\Upsilon}_p = \mathbf{I}_n$). The relevance of the second model will become apparent once we consider the time series properties of the data. The first model is the data generation process, and analyses of economic forecasting have often been based on this assumption. However, empirical econometric models are invariably not facsimiles of the process that generated the data, and in this sense are false. In place of the notion of a 'true' model, the concept of a **congruent** model has been developed as an operational criterion for evaluating models (Hendry, 1995). A model is said to be congruent if it matches the data evidence in all measurable respects – which does not require that it is either complete or correct. Key elements of congruency include that the empirical model's error is a homoscedastic innovation against the available information; that the conditioning variables are weakly exogenous for the parameters of the model (Engle, Hendry and Richard, 1983); and that those parameters are constant. In principle it is possible to design models with all of these characteristics which thereby exhaust the available information, and in essence represent the data generation process as observed so far up to an innovation error.

We suppose that the data generation process is integrated of order 1 (denoted $I(1)$), and that it satisfies $r < n$ cointegration relations such that

$$\boldsymbol{\Upsilon} = \mathbf{I}_n + \alpha\beta', \tag{3.5}$$

where α and β are $n \times r$ full rank matrices. As well as excluding explosive variables by assuming that none of the roots of $|\mathbf{I} - \boldsymbol{\Upsilon}L| = 0$ lies inside the unit circle (where L is the lag operator defined by $L^k\mathbf{x}_t = \mathbf{x}_{t-k}$), we also rule out $\mathbf{x}_t$ being integrated of order 2. Thus, we assume that $\alpha'_\perp \boldsymbol{\Theta}\beta_\perp$ is full rank, where $\boldsymbol{\Theta}$ is the mean-lag matrix (here $\boldsymbol{\Upsilon}$), and $\alpha_\perp$ and $\beta_\perp$ are full column rank $n \times (n-r)$ matrices such that $\alpha'\alpha_\perp = \beta'\beta_\perp = 0$.

Nelson and Plosser (1982) show that many economic time series can be described as $I(1)$ processes, in the sense that they contain a single unit root in the autoregressive lag polynomial of their ARMA representations. Although this has been disputed (see, for example, Perron, 1989), it serves as a reasonable assumption. Cointegration is a relationship that may hold between integrated economic time series, and implies that some of the series have common unit root components that cancel when we take appropriate linear combinations of the series. Cointegration implies restrictions, such as (3.5), on systems of equations; see, *inter alia*, Engle and Granger (1987), Johansen (1988), Phillips (1991). Banerjee *et al.* (1993) provide a general exposition of the literature on integration and cointegration.

Assuming the series are $I(1)$, (3.2) can be reparametrized as a vector equilibrium-correction model

$$\Delta\mathbf{x}_t = \tau + \alpha\beta'\mathbf{x}_{t-1} + \nu_t. \tag{3.6}$$

Both $\Delta\mathbf{x}_t$ and $\beta'\mathbf{x}_t$ are $I(0)$ but may have non-zero means. Let

$$\tau = \gamma - \alpha\mu. \tag{3.7}$$

Then we can write (3.6) as

$$(\Delta\mathbf{x}_t - \gamma) = \alpha\left(\beta'\mathbf{x}_{t-1} - \mu\right) + \nu_t \tag{3.8}$$

so that the system grows at the rate $E[\Delta\mathbf{x}_t] = \gamma$ and the long-run solution is

$$E[\beta' \mathbf{x}_t] = \mu.$$

Thus, in (3.8), both $\Delta \mathbf{x}_t$ and $\beta' \mathbf{x}_t$ are expressed as deviations about their means.

The decomposition using $\tau = \gamma - \alpha\mu$ is not orthogonal since $\gamma'\alpha\mu \neq 0$. In the Appendix we compare this decomposition with the orthogonal decomposition of Johansen (1994).

As it stands, (3.7) specifies a relation between τ, γ, α and μ. However, in deducing the causal interrelationships between the parameters the economics context is important. To a first approximation, we might assume that γ, α and μ are variation-free, while admitting the possibility that μ is a deterministic function of γ. For example, Hendry and von Ungern-Sternberg (1981) show that the equilibrium value of the savings ratio (μ) may depend on the underlying growth rate of consumption and income (γ). In general, though, the underlying growth rate, the equilibrium mean, and the speeds of adjustment to equilibrium (α) might be viewed as being determined independently by separate sets of factors. One consequence is that (τ, Υ) are not variation-free, with causality in the direction of Υ to τ. For (τ, Υ) to be variation-free when α changes, γ, α and μ could not be.

In a stationary context, we can interpret whether or not (τ, Υ) are variation-free in terms of whether the long-run mean of the process changes in response to shifts in Υ. From (3.2), the relationship between the long-run mean $\pi = E[\mathbf{x}_t]$ and τ is given by

$$\tau = (\mathbf{I}_n - \Upsilon)\pi.$$

If π remains unaltered when speeds of adjustment, Υ, change, then τ is forced to change: with a fixed long-run mean the assumption that (τ, Υ) are variation-free is untenable. Equation (3.7) clarifies directly that changes in only one of γ, α or μ must force τ to alter. In section 3.8, we show that forecasts are unconditionally unbiased when α changes but γ and μ are fixed so τ alters, but are biased when (τ, Υ) are variation free. In short, when there are structural breaks the properties of forecasts may depend crucially on which other parameters alter and which are variation-free. This is a property of the economy, and may only be discoverable by careful empirical work. In section 3.8 we analyse a number of possibilities.

Equation (3.8) is isomorphic to (3.2). The first model ((3.3) with $\tau_p = \tau$, $\Upsilon_p = \Upsilon$) was correctly specified for (3.2) and therefore also coincides with (3.8). In terms of (3.8), the second model is given by

$$\Delta \mathbf{x}_t = \gamma + \xi_t$$

which is correctly specified when $\alpha = 0$ in (3.8), in which case $\xi_t = \nu_t$. It is a vector autoregression in the differences of the variables, and is misspecified by omitting the cointegrating vectors (see the chapter by Johansen in this volume). This model is not congruent when the series cointegrate, since its error will not be an innovation against the past of the process. However, it characterizes the type of model that would result from the Box–Jenkins time series modelling tradition, where the data are differenced a sufficient number of times to remove unit roots (once, in this example) prior to modelling. It is a simple example in that the differenced series are then modelled as a constant, but little seems to be gained for our analysis by allowing for further lags in differences.

3.5 Measuring forecast accuracy

This section draws heavily on Clements and Hendry (1993a; 1993b). Measures of forecast accuracy are often based on the forecast error second moment matrix

$$\mathbf{V}_h \equiv E\left[\mathbf{e}_{T+h}\mathbf{e}'_{T+h}\right] = \text{var}\left[\mathbf{e}_{T+h}\right] + E\left[\mathbf{e}_{T+h}\right] E\left[\mathbf{e}'_{T+h}\right] \qquad (3.9)$$

where $\mathbf{e}_{T+h}$ is a vector of h-step-ahead forecast errors. This is the mean squared forecast error matrix, equal to the forecast error (co)variance matrix when forecasts are unbiased. Although in some ways a natural basis for assessing forecast accuracy, comparisons based on (3.9) may yield inconsistent rankings between forecasting models or methods for multi-step-ahead forecasts.

However, this formulation begs the question why one should consider general loss functions in the first place, and an often-heard view is that analyses should begin with the specification of the loss function, from which the optimal predictor for the problem in hand can be derived. While in some instances there may be a well-defined mapping between forecast errors and the cost of making those errors, yielding a natural measure for assessing forecast quality, this is not typically the case in macro-economics. For example, Granger and Newbold (1973, p. 38), remarked that: 'We must admit that the use of the least squares principle is dictated chiefly by convenience. However, in reality there may not exist any clearly superior alternative'. Such a view is reaffirmed by Granger (1993, p. 651): 'real-world cost functions are rarely available'.

The problem with measures based on (3.9) is that they lack invariance to nonsingular, scale-preserving linear transformations, although the model class is invariant. If, in a particular application, the focus is genuinely and exclusively on predicting a certain transformation of the data, the levels of the series, say, then it would not matter whether the rankings altered for comparisons in terms of predicting differences, say. West (1993) argues persuasively that when there is a natural measure of forecast quality, the fact that other measures give different rankings is irrelevant. In empirical practice, mean squared forecast error measures appear to hold sway, and in Monte Carlo studies of the relative forecast performance of different methods dependence of the results on the transformation of the variables on which accuracy is assessed is often a nuisance, since it precludes any generality for the findings beyond the specific transformation of the data for which the results were derived. As an example, Clements and Hendry (1993a) show that the findings of Engle and Yoo (1987), which appear to favour imposing cointegration restrictions, are a case in point. A particular method of imposing cointegration restrictions yields superior forecasts to not doing so, when comparisons are made in terms of the trace of (3.9) for predictions of the levels of the series. However, this result is not robust to predicting a scale-preserving $I(0)$ transformation of the data: the cointegrating combination of the levels of the variables and the changes of one of the series.

Clements and Hendry (1993a) show analytically that for multistep forecasts the trace of $\mathbf{V}_h$ is a measure which lacks invariance. Similarly, so also is the determinant of $\mathbf{V}_h$; and in fact, taking the matrix as a whole is insufficient to ensure invariance, as in, for example, Granger and Newbold (1977, p. 228), who choose the method/model for which $\mathbf{d}'\mathbf{V}_h\mathbf{d}$ is the smallest for every non-zero vector $\mathbf{d}$: we refer to this as the **mean squared forecast error matrix criterion.**

Formally, we can establish these and related propositions as follows. Denote a linear forecasting system by the succinct notation

$$\mathbf{\Gamma s}_t = \mathbf{u}_t \tag{3.10}$$

where $\mathbf{u}_t \sim ID_{n+k}(0, \mathbf{\Omega})$, i.e. independently distributed, zero mean with covariance matrix $\mathbf{\Omega}$, $\mathbf{s}_t' = (\mathbf{x}_t' : \mathbf{z}_t')$, $\mathbf{x}_t$ are the n variables to be forecast and $\mathbf{z}_t$ are k available predetermined variables (perhaps just $\mathbf{x}_{t-1}$) and $\mathbf{\Gamma} = (\mathbf{I} : -\mathbf{B})$ say. The parameters are $(\mathbf{B} : \mathbf{\Omega})$, where $\mathbf{\Omega}$ is symmetric, positive semi-definite. Then the likelihood and generalized variance of the system in (3.10) are invariant under scale-preserving, nonsingular transformations of the form

$$\mathbf{M\Gamma P}^{-1}\mathbf{Ps}_t = \mathbf{Mu}_t$$

or

$$\mathbf{\Gamma}^*\mathbf{s}_t^* = \mathbf{u}_t^* \quad \text{so} \quad \mathbf{u}_t^* \sim ID_{n+k}\left(0, \mathbf{M\Omega M}'\right). \tag{3.11}$$

In (3.11), $\mathbf{s}_t^* = \mathbf{Ps}_t$, $\mathbf{M}$ and $\mathbf{P}$ are respectively $n \times n$ and $(k+n) \times (k+n)$ nonsingular matrices where $\text{abs}\,(|\mathbf{M}|) = 1$ and $\mathbf{P}$ is upper block-triangular, for example

$$\mathbf{P} = \begin{bmatrix} \mathbf{I}_n & \mathbf{P}_{12} \\ 0 & \mathbf{P}_{22} \end{bmatrix} \quad \text{so that} \quad \mathbf{P}^{-1} = \begin{bmatrix} \mathbf{I}_n & -\mathbf{P}_{12}\mathbf{P}_{22}^{-1} \\ 0 & \mathbf{P}_{22}^{-1} \end{bmatrix}.$$

Since we need to be able to calculate $\mathbf{P}^{-1}$, a restriction on $\mathbf{P}$ is that $|\mathbf{P}_{22}| \neq 0$. Then

$$\mathbf{\Gamma}^* \equiv \mathbf{M\Gamma P}^{-1} = \mathbf{M}\left(\mathbf{I} : -(\mathbf{P}_{12} + \mathbf{B})\,\mathbf{P}_{22}^{-1}\right) = \mathbf{M}\left(\mathbf{I} : -\mathbf{B}^*\right). \tag{3.12}$$

No restrictions are imposed by these transforms, so the systems (3.10) and (3.11) are isomorphic. Transformations in the class shown in (3.11) are regularly undertaken in applied work, and include as special cases differences, cointegrating combinations, substitution of identities, and differentials *inter alia*. The formalization merely makes explicit the fact that a linear model is defined by its invariance under linear transformations (or affine transforms more generally). Forecasts and forecast confidence intervals made in the original system and transformed after the event to $\mathbf{x}_t^*$ or made initially from the transformed system will be identical, and this remains true when the parameters are estimated by maximum likelihood.

If we let $\mathbf{V}_h$ denote the mean squared forecast error matrix for $\mathbf{x}_t$, and $\mathbf{V}_h^*$ the mean squared forecast error matrix for $\mathbf{x}_t$ for any other method, then for transformations involving $\mathbf{M}$ only:

(*i*) the trace measure for $\mathbf{x}_t$ is not in general equivalent to that for $\mathbf{x}_t^*$

$\operatorname{tr}(\mathbf{MV}_h\mathbf{M}') \neq \operatorname{tr}(\mathbf{V}_h)$;

(*ii*) the determinant of the matrix is invariant

$|\mathbf{MV}_h\mathbf{M}'| = |\mathbf{V}_h|$ when $|\mathbf{M}| = 1$;

(*iii*) the Granger–Newbold matrix measure is invariant

$\mathbf{d}^{*\prime}\mathbf{V}_h\mathbf{d}^* < \mathbf{d}^{*\prime}\mathbf{V}_h^*\mathbf{d} \quad \forall \mathbf{d}^* \neq 0$ implies that $\mathbf{d}'\mathbf{MV}_h\mathbf{M}'\mathbf{d} < \mathbf{d}'\mathbf{MV}_h^*\mathbf{M}'d \quad \forall \mathbf{d} \neq 0$, and for all scale-preserving $\mathbf{M}$.

For transformations using $\mathbf{P}$ (as happens with the transformation from $\mathbf{x}_t$ to $\Delta\mathbf{x}_t$, say) both the determinant and the whole-matrix criteria are no longer invariant for multistep forecasts. In fact, for $h > 1$ the invariance property requires that the covariance terms between different step-ahead forecast errors be taken in to account, leading to a generalized mean squared forecast error matrix

$$\mathbf{\Phi}_h = E\left[\mathbf{E}_h\mathbf{E}_h'\right],$$

where $\mathbf{E}_h$ is obtained by stacking the forecast step errors up to and including the h-step-ahead errors

$$\mathbf{E}'_h = \left[\mathbf{e}'_{T+1}, \mathbf{e}'_{T+2}, \ldots, \mathbf{e}'_{T+h-1}, \mathbf{e}'_{T+h}\right].$$

This is known as the **generalized forecast error second moment matrix**, and comparisons of positive definiteness based on this matrix, or comparisons based on the determinant of this matrix, provide invariant measures.

For example, $|\mathbf{\Phi}_h|$ is unaffected by transforming the data by $\mathbf{M}$ (where $|\mathbf{M}| = 1$). Denote the vector of stacked forecast errors from the transformed model as $\tilde{\mathbf{E}}'_h$, so that

$$\tilde{\mathbf{E}}'_h = \left[\mathbf{e}'_{T+1}\mathbf{M}', \mathbf{e}'_{T+2}\mathbf{M}', \ldots, \mathbf{e}'_{T+h-1}\mathbf{M}', \mathbf{e}'_{T+h}\mathbf{M}'\right],$$

or $\tilde{\mathbf{E}}_h = (\mathbf{I} \otimes \mathbf{M})\,\mathbf{E}_h$. Thus

$$\begin{aligned}
\left|\tilde{\mathbf{\Phi}}_h\right| &= \left|E\left[\tilde{\mathbf{E}}_h \tilde{\mathbf{E}}'_h\right]\right| \\
&= \left|E\left[(\mathbf{I}\otimes\mathbf{M})\,\mathbf{E}_h\mathbf{E}'_h(\mathbf{I}\otimes\mathbf{M}')\right]\right| = \left|E\left[\mathbf{E}_h\mathbf{E}'_h\right]\right| \times |\mathbf{I}\otimes\mathbf{M}|^2 \\
&= \left|E\left[\mathbf{E}_h\mathbf{E}'_h\right]\right|
\end{aligned}$$

since $|\mathbf{I}\otimes\mathbf{M}| = |\mathbf{I}\otimes\mathbf{M}'| = 1$.

Transforming by $\mathbf{P}$ leaves the error process $\{\mathbf{e}_{T+i}\}$, and therefore $\mathbf{E}_h$, unaffected, demonstrating invariance to $\mathbf{P}$ transforms.

Generalizing the Granger–Newbold matrix criterion to apply to the pooled or stacked forecast error second moment matrix $\mathbf{\Phi}_h$, then the model denoted by $\sim$ dominates that denoted by $\wedge$ if

$$\hat{\mathbf{\Phi}}_h - \tilde{\mathbf{\Phi}}_h \succ 0,$$

that is, if the difference between the two estimates of the stacked forecast error second moment matrix is positive definite. It follows immediately that dominance on this measure is sufficient but not necessary for dominance on the determinant of the generalized forecast error second moment matrix

$$\hat{\mathbf{\Phi}}_h - \tilde{\mathbf{\Phi}}_h \succ 0 \Rightarrow \left|\hat{\mathbf{\Phi}}_h\right| > \left|\tilde{\mathbf{\Phi}}_h\right|,$$

since $\hat{\mathbf{\Phi}}_h$ and $\tilde{\mathbf{\Phi}}_h$ are positive definite (see, for example, Dhrymes, 1984, Proposition. 66, p. 77). Thus dominance on the determinant of the generalized forecast error second moment matrix measures is the weaker condition. It can also be related to the statistical literature on predictive likelihood; see Bjørnstad (1990) for a recent review. The predictive likelihood approach may be applicable in a wider class of model than the generalized second moment. In particular, $|\mathbf{\Phi}|$ assumes the existence of estimator second moments, and the sufficiency of a second-moment measure.

Measures other than the determinant of $\mathbf{\Phi}$ and the generalization of the Granger–Newbold criterion to the stacked matrix are possible. Since $\mathbf{\Phi}$ is symmetric

$$\mathbf{\Phi}_h = \mathbf{Q}\mathbf{\Lambda}\mathbf{Q}'$$

where $\mathbf{Q}$ is orthogonal, $\mathbf{\Lambda} = \operatorname{diag}\{\lambda_1, \lambda_2, \ldots\}$, and the λ_i are the characteristic roots of $\mathbf{\Phi}$. The order of $\mathbf{\Phi}_h$ is $h \times n$, where n is the dimension of the vector of variables. Thus

$$|\mathbf{\Phi}_h| = \prod_{i=1}^{nh} \lambda_i.$$

When $r_{\mathbf{\Phi}_h} \equiv \operatorname{rank}(\mathbf{\Phi}_h) < nh$, corresponding to $nh - r_{\mathbf{\Phi}_h}$ zeros along the diagonal of $\mathbf{\Lambda}$, the determinant is not available but other comparisons are possible, such as the maximum (or, say, minimum non-zero) root of $\mathbf{\Phi}_h$. However, such comparisons would lack invariance: the determinant seems to be the only scalar invariant. When $|\mathbf{\Phi}_h| = 0$ because of exact identities between the variables, a better solution would be to delete the redundant variables. Nevertheless, $\mathbf{\Phi}_h$ could still be singular, suggesting one should turn to comparisons based on the whole stacked matrix, $\mathbf{\Phi}$. Of course, when the relation between two $\mathbf{\Phi}_h$ matrices, $\hat{\mathbf{\Phi}}_h$ and $\tilde{\mathbf{\Phi}}_h$, say, is indefinite, some arbitrary choice of ranking seems inevitable.

To fix ideas, we end this section with a simple example. The aim is to assess the relative forecast accuracy of two models, both of which are misspecified for the process generating the data. The example originates in Baillie (1993) and Newbold (1993) who compare forecasts from an autoregressive model and a white noise model for a process which is a moving average. Their intention was to illustrate the dependence of the ranking between misspecified models on the forecast horizon, and thus warn against focusing exclusively on one-step-ahead forecasts. However, as shown in Clements and Hendry (1993b), the example can be extended to illustrate issues of (in)variance.

The data generation process is

$$y_t = \epsilon_t + \theta\epsilon_{t-1}, \quad \text{where} \quad \epsilon_t \sim IN\left(0, \sigma_\epsilon^2\right), \tag{3.13}$$

and the two rival models are

$$M_1: \quad y_t = \rho y_{t-1} + \nu_t \quad \text{where} \quad \nu_t \sim_h IN\left(0, \sigma_\nu^2\right)$$

$$M_2: \quad y_t = \eta_t \qquad\qquad \text{where} \quad \eta_t \sim_h IN\left(0, \sigma_\eta^2\right)$$

where $\sim_h$ denotes 'is hypothesized to be distributed as', so that M_1 is the first-order autoregression, and M_2 the white noise model. Matching the first-order autocorrelation from (3.13) and M_1, we deduce that $\rho = \theta / (1+\theta^2)$. The forecasts of $(y_{T+1}, \ldots, y_{T+H})$ from M_1 (denoted by ^) and M_2 (~) for known parameters and a horizon $j > 1$ are given by

	Levels	Differences
M_1	$\hat{y}_{T+j} = \rho \hat{y}_{T+j-1}$	$\Delta\hat{y}_{T+j} = \rho^{j-1}(\rho - 1) y_T;$
M_2	$\tilde{y}_{T+j} = 0$	$\Delta\tilde{y}_{T+j} = 0.$

For one-step, from M_1, $\hat{y}_{T+1} = \rho y_T$ and $\Delta\hat{y}_{T+1} = (\rho - 1) y_T$; and from M_2, $\tilde{y}_{T+1} = 0$ and $\Delta\tilde{y}_{T+1} = -y_T$.

The first point is that for one-step forecasts using mean squared forecast error on levels or differences, M_1 'wins': $M_1 : \text{var}[\hat{u}_{T+1}] = \sigma_\epsilon^2 (1+\theta^2)(1-\rho^2)$, whereas $M_2 : \text{var}[\tilde{u}_{T+1}] = \sigma_\epsilon^2 (1+\theta^2)$, where u denotes the forecast error. Next, using the same criterion for 2-step forecasts of the levels, $M_1 : \text{var}[\hat{u}_{T+2}] = \sigma_\epsilon^2 (1+\theta^2)(1+\rho^4)$, whereas $M_2 : \text{var}[\tilde{u}_{T+2}] = \sigma_\epsilon^2 (1+\theta^2)$, so M_2 'wins'.

In terms of the generalized forecast error second moment, for two-step forecasts, we have

$$M_1 : \quad \left|\hat{\mathbf{\Phi}}_2\right| = \sigma_\epsilon^2 (1+\theta)^2 \begin{bmatrix} (1-\rho)^2 & \rho \\ \rho & (1+\rho^4) \end{bmatrix}$$

and

$$M_2 : \quad \left|\tilde{\mathbf{\Phi}}_2\right| = \sigma_\epsilon^2 (1+\theta)^2 \begin{bmatrix} 1 & \rho \\ \rho & 1 \end{bmatrix},$$

so, contrary to the finding for the simple mean squared forecast error criterion, M_1 wins on this measure,

$$\left|\hat{\mathbf{\Phi}}_2\right| / \left|\tilde{\mathbf{\Phi}}_2\right| = (1-\rho^2) - \rho^6 / (1-\rho^2) < 1.$$

However, for two-step forecasts of the differences, using the mean-squared measure we can reverse the outcome relative to the two-step-ahead ranking in levels. Using v to denote a forecast error from predicting the differences of the series,

$$M_1 : \quad \text{var}[\hat{v}_{T+2}] = 2\sigma_\epsilon^2 (1+\theta^2)(1-\rho)\left[1 - \frac{1}{2}\rho^2 (1+\rho)\right]$$

whereas

$$M_2 : \quad \text{var}[\tilde{v}_{T+2}] = 2\sigma_\epsilon^2 (1+\theta^2)(1-\rho),$$

so now M_1 'wins'. Such switches of ranking can be ruled out by the use of the invariant generalized measure. In our terminology, the transformation from levels to differences is an example of a $\mathbf{P}$ transformation, for which we have already established the invariance of the generalized measure.

It is somewhat tedious to perform some of the forecast measure calculations for horizons greater than 2, so we resorted to Monte Carlo. Figure 3.1 shows the results for horizons up to 12 steps ahead, for $\theta = 0.5$ (thus $\rho = 0.4$), $\sigma_\epsilon^2 = 1$, and 10 000 replications. For horizons greater than 2 there is little to choose between the two models on the mean squared forecast error measure, whether the evaluation is in terms of the ability to predict the levels or differences of the data. However, the generalized measure implicitly weights the levels and differences forecasts, and the performance across horizons. It highlights that on average M_1 is the better model.

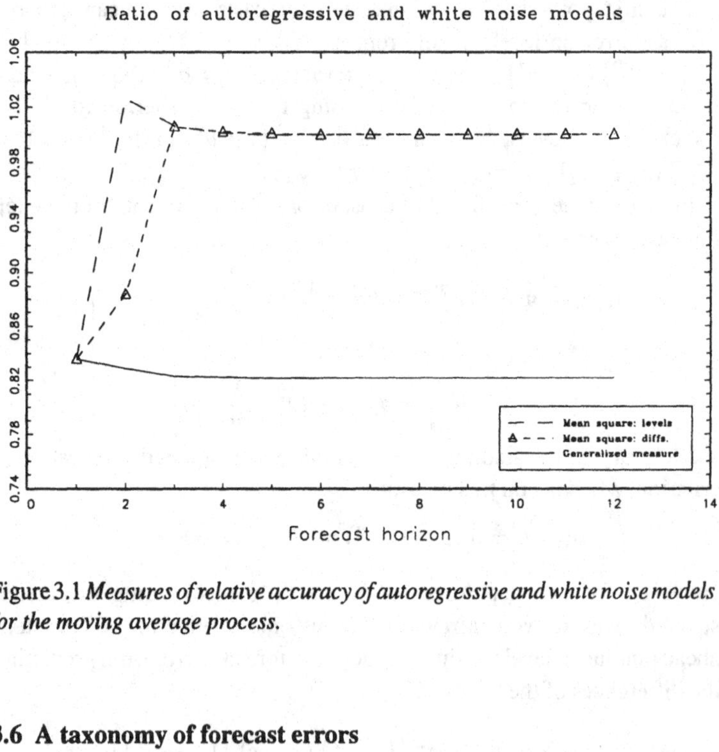

Figure 3.1 *Measures of relative accuracy of autoregressive and white noise models for the moving average process.*

3.6 A taxonomy of forecast errors

In this section, we develop a taxonomy of forecast errors. We allow a structural change in the forecast period, the model and data generation process to differ over the sample period, and recognize that the parameters of the model have to be estimated from the data, and that the forecast commences from initial conditions (denoted by $\hat{\mathbf{x}}_T$) which may differ from the 'true' values $\mathbf{x}_T$. Forecasts are often conditioned on poor provisional statistics, which are frequently subject to revisions before being finalized; see, for example, Patterson (1995). Then, j-step-ahead forecasts are given by

$$\hat{\mathbf{x}}_{T+j} = \hat{\tau} + \hat{\Upsilon}\hat{\mathbf{x}}_{T+j-1} = \sum_{i=0}^{j-1} \hat{\Upsilon}^i \hat{\tau} + \hat{\Upsilon}^j \hat{\mathbf{x}}_T, \quad j = 1, \ldots, h, \tag{3.14}$$

under the possibly mistaken assumption of parameter constancy, where '∧'s on parameters denote estimates, and on random variables, forecasts. The forecast errors are

$$\hat{\nu}_{T+j} = \mathbf{x}_{T+j} - \hat{\mathbf{x}}_{T+j}.$$

Suppose the system experiences a step change between the estimation and forecast periods, such that $(\tau : \Upsilon)$ changes to $(\tau^* : \Upsilon^*)$ over $j = 1, \ldots, h$, and the variance, autocorrelation and distribution of the error may change to $\nu_{T+j} \sim D_n(0, \Omega^*)$. More complicated changes are obviously feasible empirically, but we have not found that their analysis yields useful insights, whereas the single break does reveal certain problems. Imposing $E[\nu_{T+j}] = 0$ involves no loss of generality when $\tau^* \neq \tau$. Thus, the data actually generated by the process for the next h periods are given by

$$\begin{aligned} \mathbf{x}_{T+j} &= \tau^* + \Upsilon^* \mathbf{x}_{T+j-1} + \nu_{T+j} \\ &= \sum_{i=0}^{j-1} (\Upsilon^*)^i \tau^* + (\Upsilon^*)^j \mathbf{x}_T + \sum_{i=0}^{j-1} (\Upsilon^*)^i \nu_{T+j-i}. \end{aligned} \tag{3.15}$$

From (3.14) and (3.15), the j-step-ahead forecast error is

$$\hat{\nu}_{T+j} = \sum_{i=0}^{j-1} (\Upsilon^*)^i \tau^* - \sum_{i=0}^{j-1} \hat{\Upsilon}^i \hat{\tau} + (\Upsilon^*)^j \mathbf{x}_T - \hat{\Upsilon}^j \hat{\mathbf{x}}_T + \sum_{i=0}^{j-1} (\Upsilon^*)^i \nu_{T+j-i}. \tag{3.16}$$

We denote deviations between sample estimates and population parameters by $\delta_\tau = \hat{\tau} - \tau_p$ and $\delta_\Upsilon = \hat{\Upsilon} - \Upsilon_p$. Then, ignoring powers and cross products in the δ's

$$\sum_{i=0}^{j-1} \left[(\Upsilon^*)^i \tau^* - \hat{\Upsilon}^i \hat{\tau} \right] \approx \sum_{i=0}^{j-1} \left[(\Upsilon^*)^i \tau^* - \left(\Upsilon_p^i \tau_p + \Upsilon_p^i \delta_\tau + \mathbf{D}_i \delta_\Upsilon^\nu \right) \right] \tag{3.17}$$

where $(\cdot)^\nu$ denotes forming a vector, since

$$\hat{\Upsilon}^i = (\Upsilon_p + \delta_\Upsilon)^i \approx \Upsilon_p^i + \sum_{k=0}^{i-1} \Upsilon_p^k \delta_\Upsilon \Upsilon_p^{i-k-1} = \Upsilon_p^i + \mathbf{C}_i.$$

Therefore, on vectoring, and using $\otimes$ to denote the Kronecker product,

$$\left(\hat{\Upsilon}^i\right)^\nu \approx \left(\Upsilon_p^i\right)^\nu + \left(\sum_{k=0}^{i-1} \Upsilon_p^k \otimes \Upsilon_p^{i-k-1\prime}\right)\delta_\Upsilon^\nu,$$

and hence

$$\begin{aligned}\hat{\Upsilon}^i\hat{\tau} &\approx \left(\Upsilon_p^i + \sum_{k=0}^{i-1}\Upsilon_p^k\delta_\Upsilon\Upsilon_p^{i-k-1}\right)(\tau_p+\delta_\tau)\\ &\approx \sum_{k=0}^{i-1}\Upsilon_p^k\delta_\Upsilon\Upsilon_p^{i-k-1}\tau_p + \Upsilon_p^i\tau_p + \Upsilon_p^i\delta_\tau,\end{aligned}$$

yielding

$$\begin{aligned}\hat{\Upsilon}^i\hat{\tau} &\approx \Upsilon_p^i\tau_p + \Upsilon_p^i\delta_\tau + \left(\sum_{k=0}^{i-1}\Upsilon_p^k \otimes \tau_p'\Upsilon_p^{i-k-1\prime}\right)\delta_\Upsilon^\nu\\ &= \Upsilon_p^i\tau_p + \Upsilon_p^i\delta_\tau + \mathbf{D}_i\delta_\Upsilon^\nu.\end{aligned}$$

It is informative to allow explicitly for the model being misspecified, and hence to decompose the term $\sum_{i=0}^{j-1}[(\Upsilon^*)^i\tau^* - \hat{\Upsilon}^i\hat{\tau}]$ in (3.17) into

$$\begin{aligned}\sum_{i=0}^{j-1}\Big[&(\Upsilon^*)^i(\tau^*-\tau) + (\Upsilon^*)^i(\tau-\tau_p) + \left((\Upsilon^*)^i - \Upsilon^i\right)\tau_p\\ &+ \left(\Upsilon^i - \Upsilon_p^i\right)\tau_p - \Upsilon_p^i\delta_\tau - \mathbf{D}_i\delta_\Upsilon^\nu\Big].\end{aligned}$$

Similarly, for the terms in (3.16) multiplied by $\mathbf{x}_T$ or $\hat{\mathbf{x}}_T$, using

$$\begin{aligned}\mathbf{C}_i\mathbf{x}_T &= \sum_{k=0}^{i-1}\Upsilon_p^k\delta_\Upsilon\Upsilon_p^{i-k-1}\mathbf{x}_T\\ &= \left(\sum_{k=0}^{i-1}\Upsilon_p^k\otimes\mathbf{x}_T'\Upsilon_p^{i-k-1\prime}\right)\delta_\Upsilon^\nu\\ &= \mathbf{F}_i\delta_\Upsilon^\nu,\end{aligned}$$

and letting $(\mathbf{x}_T - \hat{\mathbf{x}}_T) = \delta_x$,

$$\begin{aligned}(\Upsilon^*)^j\mathbf{x}_T - \hat{\Upsilon}^j\hat{\mathbf{x}}_T &\approx \left((\Upsilon^*)^j - \Upsilon^j\right)\mathbf{x}_T + \left(\Upsilon^j - \Upsilon_p^j\right)\mathbf{x}_T - \mathbf{F}_i\delta_\Upsilon^\nu\\ &+ \left(\Upsilon_p^j + \mathbf{C}_j\right)\delta_x.\end{aligned}$$

There is no necessity that cross products involving δ_Υ and δ_x be negligible, so interaction terms may arise from the initial conditions.

To this order of approximation, conditional on $\mathbf{x}_T$, one taxonomy of forecast errors in (3.16) is shown in Table 3.1. At first sight, the expression

Table 3.1 *Forecast error taxonomy.*

$\hat{\nu}_{T+j} \simeq$	
$\sum_{i=0}^{j-1}\left((\Upsilon^*)^i - \Upsilon^i\right)\tau_p$	
$+\left((\Upsilon^*)^j - \Upsilon^j\right)\mathbf{x}_T$	(*i*) slope change
$+\sum_{i=0}^{j-1}(\Upsilon^*)^i(\tau^* - \tau)$	(*ii*) intercept change
$+\sum_{i=1}^{j-1}\left(\Upsilon^i - \Upsilon_p^i\right)\tau_p$	
$+\left(\Upsilon^j - \Upsilon_p^j\right)\mathbf{x}_T$	(*iii*a) slope misspecification
$+\sum_{i=0}^{j-1}(\Upsilon^*)^i(\tau - \tau_p)$	(*iii*b) intercept misspecification
$-\sum_{i=1}^{j-1}\mathbf{D}_i\delta_\Upsilon^\nu - \mathbf{F}_i\delta_\Upsilon^\nu$	(*iv*a) slope estimation
$-\sum_{i=0}^{j-1}\Upsilon_p^i\delta_\tau$	(*iv*b) intercept estimation
$+\left(\Upsilon_p^j + \mathbf{C}_j\right)\delta_x$	(*v*) initial condition
$+\sum_{i=0}^{j-1}(\Upsilon^*)^i\nu_{T+j-i}$	(*vi*) error accumulation.

is both somewhat daunting and not suggestive of positive implications. However, the decomposition supports a meaningful interpretation of the forecast error. Conditional on $\mathbf{x}_T$, the first four rows only have bias effects, whereas the remainder may also affect forecast error variances. Further, the decomposition simplifies in various states of nature, and these correspond to various elements vanishing in Table 3.1. For example, (*iiia*)–(*iiib*) vanish if the model is 'correctly specified' in sample, but otherwise the formula remains intact on replacing (Υ_p, τ_p) by (Υ, τ). Similarly, if parameters remain constant, (*i*)–(*ii*) disappear, and the formula applies with $(\Upsilon^*, \tau^*) = (\Upsilon, \tau)$. When $\mathbf{x}_T = \hat{\mathbf{x}}_T$, (*v*) vanishes. If parameter values are imposed rather than estimated, (*iva*)–(*ivb*) disappear, and so does $\mathbf{C}_j$ in (*v*). However, (*vi*) will only vanish under omniscience.

Typical analyses of forecast errors only entertain (*iv*) and (*vi*). Thus, in deriving second-moment measures of forecast accuracy, and assuming that parameter estimators are $\sqrt{T}$ consistent, the impact of estimation uncertainty appears to be of order T^{-1} and the impact of the accumulation of future disturbances dominates. Asymptotically, forecasts are only uncertain to the extent that the process generating the data is stochastic and forecasting accumulates the inherent uncertainty at each point in time.

Our primary concern will be with parameter change, so we will neglect most of the other sources of forecast error to highlight the interplay between parameter change and model misspecification. In section 3.4 we remarked that changes in the slope parameter may cause changes in the intercept, so that although (*ii*) may be considered by itself, taking (*i*) in isolation of (*ii*) may not be legitimate.

However, before pursuing that route, we briefly note how some of the other sources of forecast error may be reduced, drawing on Clements and Hendry (1994b; 1995b).

Model misspecification

Econometrics would not be viable if it required the assumption that the model was the data generation process. Econometrics is not unique in this regard, but the point is worth emphasizing since it is sometimes assumed that the model is the data generation process. Hendry (1995) presents an econometric modelling methodology based on the operational notion of congruency, as described in section 3.4. Moreover, the model is parametrized so that the included variables are relatively orthogonal: to the extent that the regressors are orthogonal to wrongly excluded influences, the parameter estimates will be consistent. Orthogonal parameter, congruent econometric models seem one of the more likely classes to deliver consistent, reliable estimates. An example is provided by the series of studies attempting to model the behaviour of aggregate consumers' expenditure in the UK. From Davidson *et al.* (1978) through Hendry, Muellbauer and Murphy (1990) to Harvey and Scott (1993), the initial dynamic and equilibrium-correction parameters recur over extended information sets and time periods, notwithstanding the considerable regime shifts which contributed to the predictive failure of the consumption function in the second half of the 1980s.

Estimation uncertainty

Estimation uncertainty is dependent on the size and information content of the sample, the quality of the model, the innovation error variance, the choice of the parametrization, and the selection of the estimator. The degree of integration of the data and whether unit roots are imposed or implicitly estimated will also interact with the above factors; see Clements and Hendry (1995a).

There is a growing literature on estimating models by minimizing (in-sample) squared multistep errors, with an early contribution due to Cox (1961) on the IMA(1,1) model. More recently, papers by Findley (1983a; 1983b) and Weiss (1991), *inter alia*, have considered AR models. The idea is that when the model is misspecified, estimation techniques based on in-sample minimization of one-step errors, e.g. ordinary least squares and maximum likelihood, may not be too bad at short leads, but could produce forecasts which go seriously awry for longer horizons. On the other hand, estimating the model by minimizing the in-sample counterpart of the desired out-of-sample forecast horizon may yield better forecasts at those horizons. Weiss (1991) shows that asymptotically in the sample size multistep estimation is optimal, in the sense of minimizing the sum of squared multistep forecast errors, given the model. However, his Monte Carlo suggests that for moderate sample sizes ordinary least squares will outperform multistep estimation. Clearly, when the model is correctly specified, multistep estimation results in a loss of efficiency.

Variable uncertainty

Initial condition uncertainty is a particular instance of variable uncertainty, and the only case to arise in closed systems.

Incorrectly measured initial values. The impact of badly measured initial values can be alleviated by improved data measurement, by supplementing the information in the latest data estimates with survey evidence, or the model's predictions for the initial period based on earlier, and hopefully more reliable, observations.

Projections of non-modelled variables. In open systems, off-line projections of non-modelled variables have to be made. This source of uncertainty can be reduced by devoting more resources to modelling the exogenous and policy variables; having a better feel for what the future will bring forth; and taking account of extraneous information on their likely evolution. Survey information also has a potentially useful role here.

Feedbacks onto non-modelled variables. Another form of variable-based forecast error arises from treating the non-modelled variables as strongly exogenous (Engle, Hendry and Richard, 1983) when it is not legitimate to do so. To support the generation of multi-step-ahead forecasts we require the absence of lagged feedback from the modelled onto the non-modelled variables. Such feedbacks would violate conditioning on future non-modelled variables, but are testable from sample information.

Invariance to policy changes. To support policy analysis, when the non-modelled variables are policy instruments, we require their super-exogeneity for the parameters of the forecasting system; see Favero and Hendry (1992) and Engle and Hendry (1993).

Error accumulation

The inherent uncertainty in the data generation process places a limit on the predictability of nonstationary dynamic mechanisms, although no empirical model is likely to attain such a bound.

3.7 Parameter constancy

The first setting we want to examine neglects regression-parameter and intercept change (*i*)–(*ii*), as well as parameter estimation (*iva*)–(*ivb*), and initial conditions mismeasurement (*v*), when (*iiia*)–(*iiib*) vanish because the model is correctly specified. Then the forecast error for the vector equilibrium-correction model is given by

$$\hat{\nu}_{T+j} = \sum_{i=0}^{j-1} \boldsymbol{\Upsilon}^i \nu_{T+j-i}. \tag{3.18}$$

Since $E[\nu_t] = 0$, the forecasts are conditionally and unconditionally unbiased

$$E[\hat{\nu}_{T+j}] = E[\hat{\nu}_{T+j} \mid \mathbf{x}_T] = 0.$$

The forecast errors from the vector autoregression in differences (denoted by '$\sim$') are derived from Table 3.1 similarly to (3.18), but noting that (*iiia*)–(*iiib*) do not vanish, but instead substituting ($\tau_p = \gamma$, $\boldsymbol{\Upsilon}_p = \mathbf{I}_n$) and rearranging

$$\begin{aligned} \tilde{\nu}_{T+j} &= \sum_{i=1}^{j-1} \left(\boldsymbol{\Upsilon}^i - \mathbf{I}_n\right) \gamma + \left(\boldsymbol{\Upsilon}^j - \mathbf{I}_n\right) \mathbf{x}_T \\ &+ \sum_{i=0}^{j-1} \boldsymbol{\Upsilon}^i (\tau - \gamma) + \sum_{i=0}^{j-1} \boldsymbol{\Upsilon}^i \nu_{T+j-i} \\ &= \sum_{i=0}^{j-1} \left(\boldsymbol{\Upsilon}^i \tau - \gamma\right) + \left(\boldsymbol{\Upsilon}^j - \mathbf{I}_n\right) \mathbf{x}_T + \sum_{i=0}^{j-1} \boldsymbol{\Upsilon}^i \nu_{T+j-i}. \end{aligned} \tag{3.19}$$

Clements and Hendry (1995a) show that $E[\tilde{\nu}_{T+j}] = 0$, so that forecasts from the vector autoregression in differences are unconditionally unbiased even though the model is misspecified.

It follows that the error in predicting $\Delta\mathbf{x}_{T+j}$, denoted by $\hat{\nu}_{\Delta,T+j}$ for the vector equilibrium-correction model is

$$\hat{\nu}_{\Delta,T+j} = \hat{\nu}_{T+j} - \hat{\nu}_{T+j-1} = \nu_{T+j} + \sum_{i=0}^{j-2} \boldsymbol{\Upsilon}^i \alpha\beta' \nu_{T+j-1-i}, \quad (3.20)$$

and similarly for the vector autoregression in differences

$$\begin{aligned} \tilde{\nu}_{\Delta,T+j} &= \tilde{\nu}_{T+j} - \tilde{\nu}_{T+j-1} \\ &= \boldsymbol{\Upsilon}^{j-1}\tau + \gamma + \left(\boldsymbol{\Upsilon}^j - \boldsymbol{\Upsilon}^{j-1}\right)\mathbf{x}_T \\ &+ \nu_{T+j} + \sum_{i=0}^{j-2} \boldsymbol{\Upsilon}^i \alpha\beta' \nu_{T+j-1-i}. \end{aligned}$$

Although they are unbiased, forecasts from the vector autoregression in differences will be less precise than those from the correctly specified model. From (3.18) and the assumption that the future disturbances are temporally uncorrelated, i.e. $E[\nu_r \nu_s'] = 0,\ r \neq s$,

$$E\left[\hat{\nu}_{T+j}\hat{\nu}_{T+j}'\right] = \mathsf{var}\left[\hat{\nu}_{T+j}\right] = \mathsf{var}\left[\hat{\nu}_{T+j} \mid \mathbf{x}_T\right] = \sum_{i=0}^{j-1} \boldsymbol{\Upsilon}^i \boldsymbol{\Omega} \boldsymbol{\Upsilon}^{i\prime}. \quad (3.21)$$

The conditional forecast error variance for the vector autoregression in differences is identical to (3.21) from (3.19), but the unconditional forecast error variance is more difficult to obtain, and from Clements and Hendry (1995a) is given by

$$\begin{aligned} \mathsf{var}\left[\tilde{\nu}_{T+j}\right] &= \sum_{r=0}^{j-1}\sum_{q=0}^{j-1} \alpha \left(\mathbf{I}_n + \beta'\alpha\right)^r \mathsf{var}\left[\mathbf{w}_{aT}\right] \left(\mathbf{I}_n + \beta'\alpha\right)^{q\prime} \alpha' \\ &+ \sum_{s=0}^{j-1} \boldsymbol{\Upsilon}^s \boldsymbol{\Omega} \boldsymbol{\Upsilon}^{s\prime} \end{aligned} \quad (3.22)$$

where

$$\mathbf{w}_{aT} = \beta'\mathbf{x}_T.$$

Comparing (3.22) to (3.21), the unconditional variance is never less than the conditional since the first term is a positive semi-definite matrix.

In terms of changes, the forecast error variance for the vector equilibrium-correction model is, from (3.20),

$$\mathsf{var}\left[\hat{\nu}_{\Delta,T+j}\right] = \mathsf{var}\left[\hat{\nu}_{\Delta,T+j} \mid \mathbf{x}_T\right] = \boldsymbol{\Omega} + \sum_{i=0}^{j-2} \boldsymbol{\Upsilon}^i \alpha\beta'\boldsymbol{\Omega}\beta\alpha'\boldsymbol{\Upsilon}^{i\prime} \quad (3.23)$$

where the last term is absent for $j = 1$, and for the vector autoregression in differences the unconditional forecast error variance is

$$\text{var}\left[\tilde{\nu}_{\Delta,T+j}\right] = \boldsymbol{\Omega} + \alpha\text{var}\left[\mathbf{w}_{aT}\right]\alpha'.$$

Letting $\boldsymbol{\Psi} = (\mathbf{I}_n + \beta'\alpha)$, the structure of $\text{var}\left[\mathbf{w}_{aT}\right]$ is such that we can write (3.23) as

$$\text{var}\left[\hat{\nu}_{\Delta,T+j}\right] = \boldsymbol{\Omega} + \alpha\text{var}\left[\mathbf{w}_{aT}\right]\alpha' - \alpha\boldsymbol{\Psi}^{j-1}\text{var}\left[\mathbf{w}_{aT}\right]\boldsymbol{\Psi}^{j-1\prime}\alpha'$$

so that the forecast error variances of the two models differ by

$$\alpha\boldsymbol{\Psi}^{j-1}\text{var}\left[\mathbf{w}_{aT}\right]\boldsymbol{\Psi}^{j-1\prime}\alpha'. \tag{3.24}$$

The difference (3.24) goes to zero in j, $(o(j))$, so that when assessing forecast accuracy using the second-moment matrix for $\Delta\mathbf{x}_t$, there is no measurable gain to using the correctly specified model beyond short forecast horizons.

This is brought out by Figure 3.2, which evaluates the asymptotic forecast error-variance formulae for both models for a bivariate system and 'typical' parameter values, where $n = 2, r = 1, \alpha = [-0.25 : 0.1]'$, $\beta = [1 : -1]'$, and $\boldsymbol{\Omega} = \mathbf{I}_2$. Each of the lines on the figure is the trace mean squared forecast error (the trace of the forecast error variance matrix, given that the forecast errors have zero mean) of the vector autoregression in differences divided by that of the vector equilibrium-correction model for assessing forecast accuracy in terms of levels ($\mathbf{x}_t$), changes ($\Delta\mathbf{x}_t$), and an $I(0)$ transformation of the data defined by $\mathbf{w}_t = [x_{1t} - x_{2t} : \Delta x_{2t}]'$. The figure emphasizes the dependence of the relative performance of the two models on the transformation of the data by which forecast accuracy is to be assessed. As discussed in section 3.5, this is a serious limitation of the trace mean squared forecast error criterion for assessing forecast accuracy, and in fact of many commonly used measures based on squared error loss more generally. However, for present purposes, notice that although generating unbiased forecasts, in the absence of parameter change the vector autoregression in differences is dominated by the vector equilibrium-correction model; any predictor has a larger mean squared error than the conditional expectation with respect to the data generation process.

3.8 Parameter non-constancy

When only the intercept changes, $\tau^* \neq \tau$, but $\boldsymbol{\Upsilon}^* = \boldsymbol{\Upsilon}$, the variance formulae derived in section 3.7 are unaltered: the mean of the forecast error (bias) may shift, but the second moment about the mean is unaffected. This case is considered in detail in Clements and Hendry (1994a), so we concentrate here on slope changes and induced

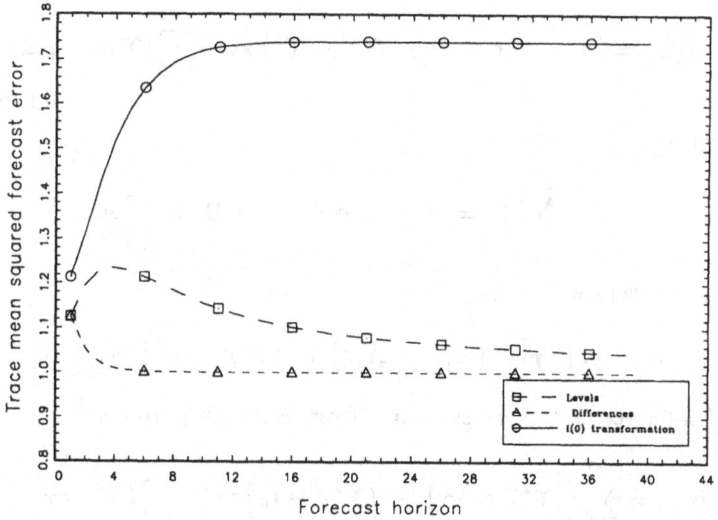

Figure 3.2 *Ratio of trace mean squared forecast error of the vector autoregression in differences to the equilibrium-correction model for horizons up to* 40 *steps ahead.*

intercept changes. The change from Υ to Υ^* could be due to changes in α, β or both. Changing α to α^* affects the dynamics around a constant equilibrium, whereas changing β to β^* alters the equilibrium and destroys previous cointegrating vectors. Thus, the latter is likely to have pernicious effects on later empirical modelling, so we focus on changes to α. Further, we assume that the changes to α are such that $\text{rank}(\alpha) = \text{rank}(\alpha^*)$, and that β is fully identified and constant, so that changes in α are not absorbed by compensating changes in β. Finally, we denote the new value of τ after α has changed to α^* by τ_π, where

$$\tau_\pi = \begin{cases} \gamma - \alpha^*\mu = \tau^* & (\tau, \Upsilon) \text{ not variation-free} \quad (i) \\ \gamma^* - \alpha^*\mu = \tau & (\tau, \Upsilon) \text{ variation-free} \quad (ii) \end{cases}$$

so that the assumption is that μ remains unaltered in both cases.

From the taxonomy of forecast errors in Table 3.1, when the slope (i) changes because α changes, with a concomitant change in τ to τ_π, the

j-step-ahead forecast error for the vector equilibrium-correction model is

$$\hat{\nu}_{T+j} = \left(\mathbf{A}_j^* \tau_\pi - \mathbf{A}_j \tau\right) + \left((\boldsymbol{\Upsilon}^*)^j - \boldsymbol{\Upsilon}^j\right) \mathbf{x}_T + \sum_{i=0}^{j-1} (\boldsymbol{\Upsilon}^*)^i \nu_{T+j-i} \tag{3.25}$$

when

$$\mathbf{A}_j = \sum_{i=0}^{j-1} \boldsymbol{\Upsilon}^i = j\mathbf{I}_n + \mathbf{B}_j \alpha \beta' \text{ where } \mathbf{B}_j = \sum_{i=0}^{j-1} \mathbf{A}_i.$$

So on average

$$E\left[\hat{\nu}_{T+j} \mid \mathbf{x}_T\right] \simeq \left(\mathbf{A}_j^* \tau_\pi - \mathbf{A}_j \tau\right) + \left((\boldsymbol{\Upsilon}^*)^j - \boldsymbol{\Upsilon}^j\right) \mathbf{x}_T. \tag{3.26}$$

For the vector autoregression in differences the j-step error is

$$\tilde{\nu}_{T+j} = \sum_{i=0}^{j-1} \left((\boldsymbol{\Upsilon}^*)^i \tau_\pi - \gamma\right) + \left((\boldsymbol{\Upsilon}^*)^j - \mathbf{I}_n\right) \mathbf{x}_T + \sum_{i=0}^{j-1} (\boldsymbol{\Upsilon}^*)^i \nu_{T+j-i} \tag{3.27}$$

and on average

$$E\left[\tilde{\nu}_{T+j} \mid \mathbf{x}_T\right] \simeq \sum_{i=0}^{j-1} \left((\boldsymbol{\Upsilon}^*)^i \tau_\pi - \gamma\right) + \left((\boldsymbol{\Upsilon}^*)^j - \mathbf{I}_n\right) \mathbf{x}_T. \tag{3.28}$$

First, consider one-step-ahead forecasts. Then

$$E\left[\hat{\nu}_{T+1} \mid \mathbf{x}_T\right] \simeq \begin{cases} (\tau^* - \tau) + (\alpha^* - \alpha)\beta'\mathbf{x}_T \\ (\alpha^* - \alpha)\beta'\mathbf{x}_T \end{cases} \tag{3.29}$$

$$= \begin{cases} (\alpha^* - \alpha)(\beta'\mathbf{x}_T - \mu) & (i) \\ (\alpha^* - \alpha)\beta'\mathbf{x}_T & (ii) \end{cases} \tag{3.30}$$

since $\tau^* - \tau = \gamma - \alpha^*\mu - (\gamma^* - \alpha^*\mu) = (\alpha - \alpha^*)\mu$ by the definition of γ^*. Thus one-step forecasts are unbiased for (*i*) when it happens that $\beta'\mathbf{x}_T = \mu$. Moreover, the constancy of the cointegrating vector β implies that $E[\beta'\mathbf{x}_T] = \mu$ and so

$$E\left[\hat{\nu}_{T+1}\right] = \begin{cases} 0 & (i) \\ (\alpha^* - \alpha)\mu & (ii) \end{cases} \tag{3.31}$$

so that vector equilibrium-correction model one-step forecasts are unconditionally unbiased under (*i*) but not (*ii*). Under (*ii*), conditional unbiasedness requires that $\beta'\mathbf{x}_T = 0$.

In terms of the vector autoregression in differences, when $j = 1$, (3.28) becomes

$$E\left[\tilde{\nu}_{T+1} \mid \mathbf{x}_T\right] \simeq \begin{cases} \left(\boldsymbol{\Upsilon}^* - \mathbf{I}_n\right)\mathbf{x}_T + \left(\tau^* - \gamma\right) & = \quad \alpha^*\left(\beta'\mathbf{x}_T - \mu\right) \quad (i) \\ \left(\boldsymbol{\Upsilon}^* - \mathbf{I}_n\right)\mathbf{x}_T + \left(\tau - \gamma\right) & \quad\;\; \alpha^*\beta'\mathbf{x}_T - \alpha\mu \quad (ii) \end{cases} \tag{3.32}$$

so that (*i*) is conditionally unbiased when the economy happens to be at equilibrium. As for the vector equilibrium-correction model, (i) is unconditionally biased, but (ii) is (conditionally and unconditionally) biased

$$E\left[\tilde{\nu}_{T+1}\right] = \begin{cases} 0 & (i) \\ \left(\alpha^* - \alpha\right)\mu & (ii) \end{cases}. \tag{3.33}$$

Now consider a long-period-ahead forecast. We need the following results, namely that for non-singular $\beta'\alpha$

$$\lim_{j\to\infty} \boldsymbol{\Upsilon}^j = \mathbf{I}_n - \alpha\left(\beta'\alpha\right)^{-1}\beta' = \mathbf{K};$$

also

$$\mathbf{K}^2 = \mathbf{K}, \;\; \mathbf{K}\alpha = 0 \text{ and } \beta'\mathbf{K} = 0;$$

and finally, $\boldsymbol{\Upsilon}^i = \mathbf{I}_n + \mathbf{A}_i\alpha\beta'$. Hence

$$\lim_{j\to\infty} \mathbf{A}_j\alpha = -\alpha\left(\beta'\alpha\right)^{-1}.$$

Similar results hold for $(\boldsymbol{\Upsilon}^*)^j$, e.g.

$$\lim_{j\to\infty} \mathbf{A}_j^*\alpha^* = -\alpha^*\left(\beta'\alpha^*\right)^{-1}.$$

Thus, for large j,

$$\mathbf{K}^* - \mathbf{K} \simeq \left(\mathbf{A}_j^*\alpha^* - \mathbf{A}_j\alpha\right)\beta'.$$

For the vector equilibrium-correction model we obtain

$$E\left[\hat{\nu}_{T+j} \mid \mathbf{x}_T\right] \to \begin{cases} \left(\mathbf{A}_j^*\tau^* - \mathbf{A}_j\tau\right) + \left(\mathbf{K}^* - \mathbf{K}\right)\mathbf{x}_T & (i) \\ \left(\mathbf{A}_j^* - \mathbf{A}_j\right)\tau + \left(\mathbf{K}^* - \mathbf{K}\right)\mathbf{x}_T & (ii) \end{cases} \tag{3.34}$$

and

$$E\left[\hat{\nu}_{T+j}\right] \to \begin{cases} 0 & (i) \\ \mathbf{A}_j^*\left(\alpha^* - \alpha\right)\mu & (ii) \end{cases}. \tag{3.35}$$

Thus, forecasts from the vector equilibrium-correction model are unbiased for this class of slope changes under (*i*).

For the vector autoregression in differences, from (3.28)

$$E\left[\tilde{\nu}_{T+j} \mid \mathbf{x}_T\right] \to \begin{cases} \mathbf{A}_j^*\alpha^*\beta'\mathbf{x}_T - \mathbf{A}_j^*\alpha^*\mu & (i) \\ \mathbf{A}_j^*\alpha^*\beta'\mathbf{x}_T - \mathbf{A}_j^*\alpha\mu & (ii) \end{cases}$$

and hence

$$E\left[\tilde{\nu}_{T+j}\right] \to \begin{cases} 0 & (i) \\ \mathbf{A}_j^*\left(\alpha^* - \alpha\right)\mu & (ii) \end{cases}$$

so that the results again match those for the vector equilibrium-correction model. Since the latter includes the cointegration feedbacks and the former does not, it is surprising that a change in the feedback coefficient has an identical effect (on average) on the forecasts of these two models, and neither is 'robust' to such a parameter change. This contrasts markedly with the outcome in response to a change in μ. Indeed, for changes to τ in response to changes in either γ or μ, Clements and Hendry (1994a) show that both models have the same non-zero bias when the forecast origin precedes the break, but that forecasts from the vector autoregression in differences made after the break has occurred are unbiased for changes in μ. For the class of break considered in this section (rank-preserving changes to α with β fixed, and with either γ or τ fixed), the timing of the break relative to the forecast origin is unimportant, and the first moment properties of the forecasts from the two models are the same. Both are unconditionally unbiased when μ and γ are fixed, i.e. τ and $\boldsymbol{\Upsilon}$ not variation-free, and otherwise are biased when γ changes.

3.9 Intercept corrections

Intercept corrections refer to the practice of specifying non-zero values for a model's error terms over the forecast period, and, as we shall show below, can be used to help robustify forecasts against structural breaks. Hendry and Clements (1994b) provide a general theory of the role of intercept corrections in macro-econometric forecasting. In the absence of the possibility of structural breaks over the future, or of other extraneous factors, it would be natural to set the future values of the equations' error terms to zero (but see below), since the underlying disturbances $\{\nu_t\}$ in (3.2) are zero mean, and most econometric estimators have the property that the estimated residuals sum to zero over the sample period.

These strictures do not apply to the vector autoregression in differences, which is misspecified when the variables cointegrate and has an error term which is correlated with past values of the process. Thus it is in part predictable, so the best guess of ξ_{t+1} in period t is not zero. The argument requires that the model's error is an innovation on the information set.

Intercept corrections are often used by the forecaster to 'adjust' the model-generated forecast in line with prior beliefs, to make an allowance for the influence of anticipated future events that are not explicitly incorporated in the specification of the model, or to 'fix up' the model for perceived misspecification over the past; see, for example, Wallis and Whitley (1991).

To illustrate the simplest form of intercept correction, suppose the forecaster reacts to perceived recent predictive failure by making an adjustment to future forecasts commensurate to the latest errors. Specifically, the forecaster adds in the residual of the current period to the next period's forecast value for one-step-ahead forecasts. Assume that the model is not misspecified and that the process remains constant. In that case the current model error is simply the realized value of the underlying disturbance term v_t in (3.36) below. The first-order autoregressive process is

$$w_t = \psi w_{t-1} + v_t, \tag{3.36}$$

where $v_t \sim IN\left(0, \sigma_v^2\right)$ and $|\psi| < 1$, and the forecast for period $T+1$ based on period T information, with $\hat{\psi}$ denoting an estimate of ψ, is

$$\hat{w}_{T+1} = \hat{\psi} w_T,$$

with a forecast error

$$e_{T+1} = w_{T+1} - \hat{w}_{T+1} = \left(\psi - \hat{\psi}\right) w_T + v_{T+1}.$$

The intercept-correcting forecast is simply

$$\hat{w}_{\iota,T+1} = \hat{w}_{T+1} + e_T = \hat{\psi} w_T + e_T,$$

with a forecast error $e_{\iota,T+1}$ given by

$$e_{\iota,T+1} = w_{T+1} - \hat{w}_{\iota,T+1} = \left(\psi - \hat{\psi}\right) w_T + v_{T+1} - e_T,$$

so that

$$e_{\iota,T+1} = \Delta e_{T+1},$$

and hence intercept correcting in this simple example differences the original forecast error. The forecast error variance will be reduced if $\mathsf{var}\left[e_{t+1}\right] = \mathsf{var}\left[e_t\right]$ and the correlation between the two residuals exceeds $\frac{1}{2}$. This example illustrates the mechanics of applying intercept corrections. As we indicate below, such adjustments are more obviously useful when the process is non-constant.

Clements and Hendry (1994a) analyse the impact of intercept corrections when there are changes in τ, corresponding to changes in the (γ, μ) parameters – the underlying growth rate and the equilibrium mean. Intercept corrections are shown largely to eliminate the bias of vector equilibrium-correction model forecasts when the break in τ has occurred before the period on which the forecasts are conditioned, although at the cost of a possibly considerable increase in forecast error variance.

We now consider intercept corrections in the presence of slope changes, of the form discussed in section 3.8, when $(\tau, \boldsymbol{\Upsilon})$ are variation-free so that γ alters. Recall that when (γ, μ) are fixed in response to a change in α, $(\tau, \boldsymbol{\Upsilon})$ can not be variation-free, and forecasts from both the vector equilibrium-correction model and the vector autoregression in differences are unconditionally unbiased. Intercept corrections are applied only to the equilibrium-correction model although a similar analysis can be conducted for the differences model. The analysis demonstrates why such corrections tend to inflate forecast error variances.

We assume that the break has occurred prior to the forecast origin, say between periods $T-1$ and T, and analyse a form of adjustment that holds the correction constant over the forecast period. The correction is again just the value of the period T residual, which for the vector equilibrium-correction model when $(\tau, \boldsymbol{\Upsilon})$ are variation-free is

$$\hat{\nu}_T = (\boldsymbol{\Upsilon}^* - \boldsymbol{\Upsilon})\,\mathbf{x}_{T-1} + \nu_T = (\alpha^* - \alpha)\,\beta'\mathbf{x}_{T-1} + \nu_T. \qquad (3.37)$$

The forecast function for the vector equilibrium-correction model with the correction given by (3.37) is obtained by solving

$$\hat{\mathbf{x}}_{\iota,T+j} = \tau + \boldsymbol{\Upsilon}\hat{\mathbf{x}}_{\iota,T+j-1} + \hat{\nu}_T$$

where $\hat{\mathbf{x}}_{\iota,T} = \mathbf{x}_T$, so that

$$\hat{\mathbf{x}}_{\iota,T+j} = \hat{\mathbf{x}}_{T+j} + \mathbf{A}_j\hat{\nu}_T = \hat{\mathbf{x}}_{T+j} + \mathbf{A}_j\,(\alpha^* - \alpha)\,\beta'\mathbf{x}_{T-1} + \mathbf{A}_j\nu_T.$$

The relationship between the forecast errors for the intercept-correcting strategy and the conditional expectation is simply

$$\hat{\nu}_{\iota,T+j} = \mathbf{x}_{T+j} - \hat{\mathbf{x}}_{\iota,T+j} = \hat{\nu}_{T+j} - \mathbf{A}_j\hat{\nu}_T \qquad (3.38)$$

where $\hat{\nu}_{T+j}$ is given by (3.25). Then, the expected value of the conditional forecast error for large j is

$$\begin{aligned} E\left[\hat{\nu}_{\iota,T+j} \mid \mathbf{x}_T\right] &\simeq \left\{\left(\mathbf{A}_j^* - \mathbf{A}_j\right)\tau + (\mathbf{K}^* - \mathbf{K})\,\mathbf{x}_T\right\} \\ &- \left(\mathbf{A}_j\,(\alpha^* - \alpha)\,\beta'\mathbf{x}_{T-1} + \mathbf{A}_j\nu_T\right), \end{aligned}$$

where the first term is (3.34) (*ii*), and unconditionally

$$E\left[\hat{\nu}_{\iota,T+j}\right] = \left(\mathbf{A}_j^* - \mathbf{A}_j\right)(\alpha^* - \alpha)\,\mu. \qquad (3.39)$$

A comparison of (3.39) and (3.35) (*ii*) indicates that the bias is reduced but not removed. However, notice that for $j = 1$, $\mathbf{A}_1^* = \mathbf{A}_1 = \mathbf{I}$, so that the correction strategy yields unbiased one-step-ahead forecasts, $E\left[\hat{\nu}_{\iota,T+1}\right] = \mathbf{0}$, eliminating a bias of $(\alpha^* - \alpha)\,\mu$.

While intercept corrections generally reduce forecast bias, there is typically a cost in terms of forecast error variance. As a baseline for the intercept-correcting strategy, we first derive the forecast error variance for the equilibrium-correction model when there is structural change, as in section 3.8. From (3.25) and (3.26)

$$\hat{\nu}_{T+j} - E\left[\hat{\nu}_{T+j} \mid \mathbf{x}_T\right] = \sum_{i=0}^{j-1} \left(\boldsymbol{\Upsilon}^*\right)^i \nu_{T+j-i},$$

so that the conditional forecast error variance is

$$\mathrm{var}\left[\hat{\nu}_{T+j} \mid \mathbf{x}_T\right] = \sum_{i=0}^{j-1} \left(\boldsymbol{\Upsilon}^*\right)^i \boldsymbol{\Omega} \left(\boldsymbol{\Upsilon}^*\right)^{i\prime} \quad (3.40)$$

irrespective of whether γ or τ is assumed fixed. This is because if γ is fixed, τ changes, and if τ is fixed, γ changes. Either way, changes to the deterministic components will not affect forecast error variances.

More interestingly, consider the unconditional forecast error variances. When γ is fixed, so $\tau_\pi = \tau^*$, then for large j (3.25) is

$$\hat{\nu}_{T+j} \simeq \left(\mathbf{A}_j^* \alpha^* - \mathbf{A}_j \alpha\right) \left(\beta' \mathbf{x}_T - \mu\right) + \sum_{i=0}^{j-1} \left(\boldsymbol{\Upsilon}^*\right)^i \nu_{T+j-i},$$

$$E\left[\hat{\nu}_{T+j}\right] = 0, \quad (3.41)$$

so that

$$\begin{aligned} \mathrm{var}\left[\hat{\nu}_{T+j}\right] &= \sum_{i=0}^{j-1} \left(\boldsymbol{\Upsilon}^*\right)^i \boldsymbol{\Omega} \left(\boldsymbol{\Upsilon}^*\right)^{i\prime} \\ &+ \left(\mathbf{A}_j^* \alpha^* - \mathbf{A}_j \alpha\right) \mathsf{V}\left[\beta' \mathbf{x}_T\right] \left(\mathbf{A}_j^* \alpha^* - \mathbf{A}_j \alpha\right)' \end{aligned} \quad (3.42)$$

Equation (3.42) also holds when $\tau_\pi = \tau$.

Consider now the impact of intercept-correcting the equilibrium-correction model when τ is fixed. From (3.38)

$$\hat{\nu}_{\iota,T+j} - E\left[\hat{\nu}_{\iota,T+j}\right] = \hat{\nu}_{T+j} - E\left[\hat{\nu}_{T+j}\right] - \mathbf{A}_j\left[\hat{\nu}_T - E\left[\hat{\nu}_T\right]\right]. \quad (3.43)$$

For large j, (3.43) is

$$\begin{aligned} &\left(\mathbf{A}_j^* \alpha^* - \mathbf{A}_j \alpha\right) \left(\beta' \mathbf{x}_T - \mu\right) + \\ &\sum_{i=0}^{j-1} \left(\boldsymbol{\Upsilon}^*\right)^i \nu_{T+j-i} - \mathbf{A}_j\left[\left(\alpha^* - \alpha\right)\left(\beta' \mathbf{x}_{T-1} - \mu\right) + \nu_T\right], \end{aligned} \quad (3.44)$$

and thus

$$\begin{aligned} \mathrm{var}\left[\hat{\nu}_{\iota,T+j}\right] &= \mathrm{var}\left[\hat{\nu}_{T+j}\right] + \mathbf{A}_j \boldsymbol{\Omega} \mathbf{A}_j' \quad (3.45) \\ &+ \mathbf{A}_j\left(\alpha^* - \alpha\right) \mathsf{V}\left[\beta' \mathbf{x}_{T-1}\right] \left(\mathbf{A}_j\left(\alpha^* - \alpha\right)\right)' - C\left(\cdot\right) \end{aligned}$$

where $C(\cdot)$ is the cross product of the first and third terms in (3.44). Relative to not intercept-correcting, there are two additional positive definite matrices in (3.45), the second and third terms, so that the forecast error variance should increase.

Additional insight can be gleaned by supposing that there are no structural breaks. Then the second term in (3.42) disappears and in the first $\boldsymbol{\Upsilon}^*$ is replaced by $\boldsymbol{\Upsilon}$. The unconditional corresponds to the conditional variance (3.40), and we are back in the constant-parameter world of section 3.7. But the impact of intercept-correcting, shown in (3.45), is to inflate the forecast error variance by $\mathbf{A}_j\boldsymbol{\Omega}\mathbf{A}_j'$ since the third and fourth terms in (3.45) disappear. This term arises because the correction term in (3.37) includes the realized value of the period T disturbance, which has a variance of $\boldsymbol{\Omega}$. More precise estimates of the required correction term could be obtained by averaging a number of recent errors, provided the break occurred that far back, and in practice this is often done. However, if the forecaster mistakenly averages over periods prior to the break having occurred then the correction term will be underestimated and the impact on the bias diluted.

An empirical example in Clements and Hendry (1994a) illustrates some of these ideas. Various forms of correction are applied to a quarterly vector equilibrium-correction model of the relationships between wages, prices and unemployment, which is believed to have altered over the 1980s. Forecast biases are reduced, but only at the cost of larger unconditional forecast error variances as the horizon increases, in agreement with the analysis presented in this section. Testing for the importance of the break prior to using intercept corrections merits consideration.

3.10 Conclusions

There are many possible sources of forecast error beyond the innovations to the data generation process and the parameter-estimation uncertainty which are reflected in conventional forecast error-variance formulae. When the data generation process is nonstationary, the model differs in important respects from the mechanism, and the process is subject to structural breaks, then mechanistic application of an estimated macro-econometric model is unlikely to yield the best forecasts. We considered the role of intercept corrections, and showed that for a class of changes to the parameters they can reduce forecast biases, at some cost in forecast error variances. A statistical testing theory for when to incorporate recent residuals merits development, and is the subject of our ongoing research. Different model forms show varying robustness to regime shifts, which is information that might be exploited as well. When models are misspecified, alternative estimators (e.g., multistep estimation) may offer some protection, particularly when the mistake concerns omitting long-memory components.

Appendix: An alternative decomposition of the vector of constant terms

The decomposition of τ adopted in the text,

$$\tau = \gamma - \alpha\mu,$$

can be contrasted to

$$\tau = \alpha_{\perp}\zeta - \alpha\beta_0,$$

where $\beta_0 = -(\alpha'\alpha)^{-1}\alpha'\tau$, and $\zeta = (\alpha'_{\perp}\alpha_{\perp})^{-1}\alpha'_{\perp}\tau$. Then $\alpha\beta_0$ and $\alpha_{\perp}\zeta$ are orthogonal by construction. Conversely, the cost of orthogonality is that, in (3.46) below, the cointegrating vectors are no longer deviations about their means and so the 'intercept' $\alpha_{\perp}\zeta_0$ is not simply the growth rate

$$(\Delta\mathbf{x}_t - \alpha_{\perp}\zeta) = \alpha(\beta'\mathbf{x}_{t-1} - \beta_0) + \nu_t. \tag{3.46}$$

From (3.46), setting $E[\Delta\mathbf{x}_t] = \gamma$

$$\alpha E[\beta'\mathbf{x}_t] = \gamma - \alpha_{\perp}\zeta + \alpha\beta_0$$

and when $\beta'\alpha$ is non-singular

$$\mu \equiv E[\beta'\mathbf{x}_t] = (\beta'\alpha)^{-1}\beta'(\gamma - \alpha_{\perp}\zeta + \alpha\beta_0) \neq \beta_0.$$

From

$$\alpha_{\perp}\zeta - \alpha\beta_0 = \gamma - \alpha\mu,$$

we obtain

$$\alpha_{\perp}\zeta = \gamma - \alpha\left[(\beta'\alpha)^{-1}\beta'(\gamma - \alpha_{\perp}\zeta)\right],$$

or

$$\gamma = \mathbf{K}\alpha_{\perp}\zeta + \alpha(\beta'\alpha)^{-1}\beta'\gamma,$$

where $\mathbf{K} = \mathbf{I}_n - \alpha(\beta'\alpha)^{-1}\beta'$. Since $\beta'\gamma = 0$ from pre-multiplying $E[\Delta\mathbf{x}_t] = \gamma$ by β' (the means of the cointegrating vectors are constant), the relation between γ, μ and the orthogonal decomposition is

$$\begin{aligned}\mu &= -(\beta'\alpha)^{-1}\beta'\alpha_{\perp}\zeta + \beta_0\\ \gamma &= \mathbf{K}\alpha_{\perp}\zeta.\end{aligned}$$

The decomposition into γ and μ is notationally simpler, and is adopted in the text.

Acknowledgments

Financial support from the UK Economic and Social Research Council under grant R000233447 is gratefully acknowledged by both authors. We are also grateful to David Cox and David Hinkley for comments on an earlier version of this paper.

References

Baillie, R.T. (1993) Comment on 'On the limitations of comparing mean squared forecast errors', by M.P. Clements and D.F. Hendry. *Journal of Forecasting*, **12**, 639–641.

Banerjee, A., Dolado, J.J., Galbraith, J.W. and Hendry, D.F. (1993) *Co-integration, Error Correction and the Econometric Analysis of Non-stationary Data*. Oxford University Press, Oxford.

Bjørnstad, J.F. (1990). Predictive likelihood: a review. *Statistical Science*, **5**, 242–265.

Box, G.E.P. and Jenkins, G.M. (1976) *Time Series Analysis, Forecasting and Control*. Holden-Day, San Francisco.

Brown, B.Y. and Mariano, R.S. (1984) Residual based stochastic predictors and estimation in nonlinear models. *Econometrica*, **52**, 321–343.

Brown, B.Y. and Mariano, R.S. (1989) Predictors in dynamic nonlinear models: large sample behaviour. *Econometric Theory*, **5**, 430–452.

Chatfield, C. (1993) Calculating interval forecasts. *Journal of Business and Economic Statistics*, **11**, 121–135.

Clements, M.P. and Hendry, D.F. (1993a) On the limitations of comparing mean squared forecast errors. *Journal of Forecasting*, **12**, 617–637. With discussion.

Clements, M.P. and Hendry, D.F. (1993b) On the limitations of comparing mean squared forecast errors: a reply. *Journal of Forecasting*, **12**, 669–676.

Clements, M.P. and Hendry, D.F. (1994a) Intercept corrections and structural change. Mimeo. Institute of Economics and Statistics, University of Oxford.

Clements, M.P. and Hendry, D.F. (1994b) Towards a theory of economic forecasting. In C. Hargreaves (ed.), *Non-stationary Time-series Analyses and Cointegration*, pp. 9–52. Oxford University Press, Oxford.

Clements, M.P. and Hendry, D.F. (1995a) Forecasting in cointegrated systems. *Journal of Applied Econometrics*, **10**, 127–146.

Clements, M.P. and Hendry, D.F. (1995b) *A Theory of Economic Forecasting*. Cambridge University Press, Cambridge.

Cook, S. and Hendry, D.F. (1993) The theory of reduction in econometrics. *Poznań Studies in the Philosophy of the Sciences and the Humanities*, **38**, 71–100.

Cox, D.R. (1961) Prediction by exponentially weighted moving averages and related methods. *Journal of the Royal Statistical Society, Series B*, **23**, 414–422.

Cox, D.R. and Miller, H.D. (1965) *The Theory of Stochastic Processes*. Chapman & Hall, London.

Davidson, J.E.H., Hendry, D.F., Srba, F. and Yeo, S. (1978) Econometric modelling of the aggregate time series relationship between consumers' expenditure and income in the United Kingdom. *Economic Journal*, **88**, 661–692. Reprinted in D.F. Hendry (1993) *Econometrics: Alchemy or Science?* Blackwell Publishers, Oxford.

Dhrymes, P.J. (1984) *Mathematics for Econometrics* (2nd edn). Springer-Verlag, New York.

Doan, T., Litterman, R. and Sims, C.A. (1984) Forecasting and conditional projection using realistic prior distributions. *Econometric Reviews*, **3**, 1–100.

Emerson, R.A. and Hendry, D.F. (1994) An evaluation of forecasting using leading indicators. Mimeo. Nuffield College, Oxford.

Engle, R.F. and Granger, C.W.J. (1987) Cointegration and error correction: representation, estimation and testing. *Econometrica*, **55**, 251–276.

Engle, R.F. and Hendry, D.F. (1993) Testing super exogeneity and invariance in regression models. *Journal of Econometrics*, **56**, 119–139.

Engle, R.F, Hendry, D.F. and Richard, J.-F. (1983) Exogeneity. *Econometrica*, **51**, 277–304. Reprinted in D.F. Hendry (1993) *Econometrics: Alchemy or Science?* Blackwell Publishers, Oxford.

Engle, R.F. and Yoo, B.S. (1987) Forecasting and testing in co-integrated systems. *Journal of Econometrics*, **35**, 143–159.

Ericsson, N.R., Campos, J. and Tran, H.-A. (1990) PC-GIVE and David Hendry's econometric methodology. *Revista de Econometria*, **10**, 7–117.

Fama, E.F. and French, K.R. (1988) Permanent and temporary components of stock prices. *Journal of Political Economy*, **96**, 246–273.

Favero, C. and Hendry, D.F. (1992) Testing the Lucas critique: a review. *Econometric Reviews*, **11**, 265–306.

Fildes, R. and Makridakis, S. (1994) The impact of empirical accuracy studies on time series analysis and forecasting. Discussion paper: Management School, Lancaster University.

Findley, D.F. (1983a) Model selection for multi-step-ahead forecasting. *Proceedings of the American Statistical Association, Business and Economic Statistics Section*, 243–247.

Findley, D.F. (1983b) On the use of multiple models for multi-period forecasting. *Proceedings of the American Statistical Association, Business and Economic Statistics Section*, 528–531.

Florens, J-P., Mouchart, M. and Rolin, J-M. (1990) *Elements of Bayesian Statistics*. Marcel Dekker, New York.

Gilbert, C.L. (1986) Professor Hendry's methodology. *Oxford Bulletin of Economics and Statistics*, **48**, 283–307. In C.W.J. Granger (ed.) (1990) *Modelling Economic Series*. Clarendon Press, Oxford.

Granger, C.W.J. (1993) Comment on 'On the limitations of comparing mean squared forecast errors', by M.P. Clements and D.F. Hendry. *Journal of Forecasting*, **12**, 651–652.

Granger, C.W.J. and Deutsch, M. (1992) Comments on the evaluation of policy models. *Journal of Policy Modeling*, **14**, 497–516.

Granger, C.W.J. and Newbold, P. (1973) Some comments on the evaluation of economic forecasts. *Applied Economics*, **5**, 35–47.

Granger, C.W.J. and Newbold, P. (1977) *Forecasting Economic Time Series*. Academic Press, New York.

Granger, C.W.J. and Teräsvirta, T. (1993) *Modelling Nonlinear Economic Relationships*. Oxford University Press, Oxford.

Haavelmo, T. (1944) The probability approach in econometrics. *Econometrica*, **12**, 1–118. Supplement.

Harvey, A.C. (1989) *Forecasting, Structural Time Series Models and the Kalman Filter*. Cambridge University Press, Cambridge.

Harvey, A.C. and Scott, A. (1993) Seasonality in dynamic regression models. Unpublished paper: London School of Economics.

Hendry, D.F. (1987) Econometric methodology: a personal perspective. In T.F. Bewley, (ed.), *Advances in Econometrics*. Cambridge University Press, Cambridge.

Hendry, D.F. (1993) The roles of economic theory and econometrics in time series economics. Invited address: European Econometric Society, Stockholm.

Hendry, D.F. (1995). *Dynamic Econometrics*. Oxford University Press, Oxford.

Hendry, D.F. and Clements, M.P. (1994a) Can econometrics improve economic forecasting? *Swiss Journal of Economics and Statistics*, **130**, 267–298.

Hendry, D.F. and Clements, M.P. (1994b) On a theory of intercept corrections in macro-economic forecasting. In S. Holly, (ed.), *Money, Inflation and Employment: Essays in Honour of James Ball*. Edward Elgar, Aldershot.

Hendry, D.F. and Mizon, G.E. (1990) Procrustean econometrics: or the stretching and squeezing of data. In C.W.J. Granger (ed.), *Modelling Economic Series*. Clarendon Press, Oxford.

Hendry, D.F. and Mizon, G.E. (1993) Evaluating dynamic econometric models by encompassing the VAR. In P.C.B. Phillips (ed.), *Models, Methods and Applications of Econometrics*. Basil Blackwell, Oxford.

Hendry, D.F. and Richard, J.-F. (1982) On the formulation of empirical models in dynamic econometrics. *Journal of Econometrics*, **20**, 3–33. Reprinted in C.W.J. Granger (ed.) (1990) *Modelling Economic Series*. Clarendon Press, Oxford; and in D.F. Hendry (1993) *Econometrics: Alchemy or Science?* Blackwell Publishers, Oxford.

Hendry, D.F. and Richard, J.-F. (1983) The econometric analysis of economic time series (with discussion). *International Statistical Review*, **51**, 111–163. Reprinted in D.F. Hendry (1993) *Econometrics: Alchemy or Science?* Blackwell Publishers, Oxford.

Hendry, D.F. and von Ungern-Sternberg, T. (1981). Liquidity and inflation effects on consumers' expenditure. In A.S. Deaton, (ed.), *Essays in the Theory and Measurement of Consumers' Behaviour*. Cambridge University Press, Cambridge. Reprinted in D.F. Hendry (1993), *Econometrics: Alchemy or Science?* Blackwell Publishers, Oxford.

Hendry, D.F., Muellbauer, J.N.J. and Murphy, T.A. (1990) The Econometrics of DHSY. In J.D. Hey and D. Winch, (eds), *A Century of Economics*. Basil Blackwell, Oxford.

Holt, C.C. (1957) Forecasting seasonals and trends by exponentially weighted moving averages. ONR Research Memorandum 52. Carnegie Institute of Technology, Pittsburgh, PA.

Johansen, S. (1988) Statistical analysis of cointegration vectors. *Journal of Economic Dynamics and Control*, **12**, 231–254.

Johansen, S. (1994) The role of the constant and linear terms in cointegration analysis of nonstationary variables. *Econometric Reviews*, **13**, 205–229.

Kalman, R.E. (1960) A new approach to linear filtering and prediction problems. *Journal of Basic Engineering*, **82**, 35–45.

Kendall, M.G., Stuart, A. and Ord, J.K. (1987) *Advanced Theory of Statistics* (5th edn), Vol. 1. Charles Griffin and Co., London.

Klein, L.R. (1971) *An Essay on the Theory of Economic Prediction*. Markham Publishing Company, Chicago.

Lahiri, K. and Moore, G.H. (eds) (1991) *Leading Economic Indicators: New Approaches and Forecasting Records*. Cambridge University Press, Cambridge.

Mariano, R.S. and Brown, B.W. (1983) Asymptotic behaviour of predictors in a nonlinear simultaneous system. *International Economic Review*, **24**, 523–536.

Mariano, R.S. and Brown, B.W. (1991) Stochastic-simulation tests of nonlinear econometric models. In L.R. Klein, (ed.), *Comparative Performance of U.S. Econometric Models*. Oxford University Press, Oxford.

Nelson, C.R. and Plosser, C.I. (1982) Trends and random walks in macro-economic time series: some evidence and implications. *Journal of Monetary Economics*, **10**, 139–162.

Newbold, P. (1993) Comment on 'On the limitations of comparing mean squared forecast errors', by M.P. Clements and D.F. Hendry. *Journal of Forecasting*, **12**, 658–660.

Patterson, K.D. (1995) An integrated model of the data measurement and data generation processes with an application to consumers' expenditure. *Economic Journal*, **105**, 54–76.

Perron, P. (1989) The great crash, the oil price shock and the unit root hypothesis. *Econometrica*, **57**, 1361–1401.

Phillips, P.C.B. (1991) Optimal inference in cointegrated systems. *Econometrica*, **59**, 283–306.

Salmon, M. and Wallis, K.F. (1982) Model validation and forecast comparisons: theoretical and practical considerations. In G.C. Chow and P. Corsi (eds), *Evaluating the Reliability of Macro-Economic Models*. John Wiley, New York.

Spanos, A. (1986) *Statistical Foundations of Econometric Modelling*. Cambridge University Press, Cambridge.

Stock, J.H. and Watson, M.W. (1989) New indexes of coincident and leading economic indicators. *NBER Macro-Economic Annual*, 351–409.

Stock, J.H. and Watson, M.W. (1992) A procedure for predicting recessions with leading indicators: econometric issues and recent experience. Working Paper 4014, NBER.

Wallis, K.F. (1989) Macroeconomic forecasting: a survey. *Economic Journal*, **99**, 28–61.

Wallis, K.F. and Whitley, J.D. (1991) Sources of error in forecasts and expectations: U.K. economic models 1984–8. *Journal of Forecasting*, **10**, 231–253.

Weiss, A.A. (1991) Multi-step estimation and forecasting in dynamic models. *Journal of Econometrics*, **48**, 135–149.

West, K.D. (1993) Comment on 'On the limitations of comparing mean squared forecast errors', by M.P. Clements and D.F. Hendry. *Journal of Forecasting*, **12**, 666–667.

Winters, P.R. (1960) Forecasting sales by exponentially weighted moving averages. *Management Science*, **6**, 324–342.

Wold, H.O.A. (1938) *A Study in the Analysis of Stationary Time Series*. Almqvist & Wiksell, Stockholm.

CHAPTER 4

Longitudinal panel data: an overview of current methodology

Nan M. Laird

4.1 Introduction

This paper presents an overview of current methodology for the analysis of longitudinal data. We characterize the general class of longitudinal studies as those where one measures a specified endpoint repeatedly on the same set of subjects over time, with the objective of studying both the level and change in outcome over time as a function of subject characteristics. We review the use of marginal regression models to parametrize the effects in longitudinal data settings, then discuss two popular methods for parameter estimation: generalized estimating equations (GEE) and maximum likelihood (ML). The ML approach uses a multivariate generalization of the univariate exponential family model that forms the basis for generalized linear regression models. Both estimation methods are introduced assuming all subjects are measured on the same T occasions; we then discuss modifications necessary to accommodate missing data, and give an example which applies both approaches to a longitudinal study of obesity in children. Finally, we review the analysis of a special class of models for longitudinal data called two-stage random effects models, and discuss their relationship to marginal regression models. We close with a brief discussion of current work dealing with drop-outs in longitudinal studies.

4.2 General formulation

Longitudinal studies are increasingly common in many areas of research including the behavioural, social, economic, health and medical sciences. The distinguishing feature of a longitudinal panel study is that the outcome of interest is measured repeatedly over time in the same subjects, with the general objective of characterizing change in the outcome over time, and studying factors which contribute to the mean level and to the change. An especially attractive feature is that because the same subjects are measured repeatedly, one has the opportunity to study intra-individual change; this same feature means that repeat responses are correlated. The correlation must be taken into account when analysing the data, in order both to gain efficiency in the analysis and to account properly for sampling variability.

In this paper we will deal with longitudinal studies where the number of subjects, say N, is generally large relative to the number of time points, say T, and the objective of the analysis is to characterize the mean response on the ith subject at the jth occasion, say $\mu_{ij} = E(Y_{ij})$, as a function of time, as well as covariates thought to influence mean response, the interaction of time and those covariates. In order to fix ideas, we consider a few examples. In the asthma studies conducted by the United States Environmental Protection Agency, asthmatics kept daily records on the presence or absence of an asthma attack on each day (the response Y_{ij}) for a series of days. Simultaneously, daily recordings were made of air quality levels. The objective was to relate air quality levels to the frequency of asthma attacks. Since we are interested in relating the rate of asthma attacks to air quality levels, a regression model for $E(Y_{ij})$ on daily air quality levels is appropriate. To take a second example, consider a clinical trial comparing the effect of a new anti-hypertensive medication to that of a standard medication on long-term changes in blood pressure. Here the main predictors are time since initiating therapy, treatment assignment and their interaction. A third example, described in Waternaux, Laird and Ware (1989), involves a study of the effect of exposure to environmental lead on developmental outcomes in infants. A sample of children were classified into three groups (high, medium, low exposure) according to the level of lead found in blood samples taken from their umbilical cord at birth. At ages 6, 12 and 18 months, the cognitive development of each child was assessed using the Mental Development Index (MDI) taken from the Bayley Scale of Infant Mental Development. In addition, blood samples drawn from the infants on those occasions were assayed for lead. The primary objectives of the study included assessing the effect of initial (cord blood) lead exposure and subsequent (infant blood) lead exposure on both the level and the rate of change in MDI. The predictors included age effects, both initial and subsequent blood lead levels, and their interactions.

We shall refer to models for the mean response as **longitudinal regression models**, where μ_{ij} will be linked to a vector of regressors, say $\mathbf{x}_{ij}$, via a known link function. For example, when the responses are continuous, the identity link $\mu_{ij} = \mathbf{x}_{ij}^T\beta$ is commonly used, whereas if Y_{ij} is binary the logit link, $\text{logit}(\mu_{ij}) = \mathbf{x}_{ij}^T\beta$, is a natural choice. These regression models are also sometimes termed **marginal models**, or **marginal regression models**, to make the point that the mean response being modelled, μ_{ij}, is conditional only on $\mathbf{x}_{ij}$, and not on values of the response obtained at some earlier occasion. The covariates included in $\mathbf{x}_{ij}$ are either non-time-varying, for example, individual characteristics such as race, sex or treatment group; non-stochastic, but time-varying, such as time, or 'condition' which may be assigned by the investigators as in cross-over studies; or time-varying and stochastic but 'external' in the sense that their values are not influenced by the stochastic outcomes, Y_{ij}.

To continue with our examples, in the asthma studies, the covariate vector includes daily measures of air quality and perhaps other factors such as temperature and humidity, and individual characteristics, including initial age, ethnicity or occupation. The variable air quality is time-varying and stochastic, but clearly external, as are other potential time-varying covariates such as temperature, humidity, etc. Individual characteristics such as initial age, ethnicity and occupation may be used to identify especially vulnerable sub-groups, as well as control for overall rate of attacks. A traditional time series analysis of these data would involve the use of history of asthma attacks (previous Y_{ij}) to predict current and future attacks. In our setting, the time series nature of the data is incidental to the main objective of the study, and regarded as a nuisance feature of the model.

In the second example, treatment assignment is fixed, and time since initiating therapy and its interaction with treatment assignment are time-varying but not stochastic. However, complications may arise in the analysis. If we achieve complete follow-up on all subjects throughout the period of the trial (meaning all blood pressure measurements are obtained for all subjects), then an intention-to-treat analysis of trial results is straightforward. Suppose, however, we know that patients may discontinue therapy during the trial, and we can in fact measure compliance to assigned medications throughout the study. Many clinicians, as well as statistical analysts, argue that additional analyses conditioning on measures of compliance can enhance our understanding of the efficacy of treatments actually received. However, compliance also may be influenced by previous outcome history, and thus is not an external predictor.

In the lead study, the predictors include age effects, initial exposure, and child's blood lead level at six months, and a time-varying variable constructed by cumulating values of blood lead concentrations. In this case it might be argued that the predictors based on child's blood lead are internal if families whose children were doing poorly (low values of Y) were likely to take steps to reduce future levels of lead exposure, X.

When we deal only with external covariates we can write

$$E(Y_{ij}|\mathbf{x}_{ij}) = E(Y_{ij}|\mathbf{X}_i), \tag{4.1}$$

where $\mathbf{X}_i$ denotes the complete set of covariates for all T time points, say

$$\mathbf{X}_i = \begin{pmatrix} \mathbf{x}_{i1}^T \\ \mathbf{x}_{i2}^T \\ \vdots \\ \mathbf{x}_{iT}^T \end{pmatrix}$$

Note that this is not meant to imply that we regress today's outcomes (asthma) on future covariates (air quality), since $E(Y_{ij})$ depends only on $\mathbf{x}_{ij}$. Rather, the implication is that in taking expectations over the random outcomes, both past and future covariate values can be regarded as fixed; this would not be the case if future covariates were determined by current responses. If $g(\cdot)$ denotes the link function, we may interchangeably write

$$g(\mu_{ij}) = \mathbf{x}_{ij}^T\beta, \quad j = 1, \ldots, T, \tag{4.2}$$

or

$$\mathbf{g}(\boldsymbol{\mu}_i) = \mathbf{X}_i\beta, \tag{4.3}$$

where $\mathbf{g}(\boldsymbol{\mu}_i)^T = (g(\mu_{i1}), g(\mu_{i2}), \ldots, g(\mu_{iT}))$ and $\mu_i^T = (\mu_{i1}, \mu_{i2}, \ldots, \mu_{iT})$ is the vector of means on the T occasions. If $\mathbf{x}_{ij}$ contains stochastic 'internal' covariates, equation (4.1) no longer holds because future values of the covariate may give information about current values of Y. In this paper we permit $\mathbf{x}_{ij}$ to contain only non-stochastic or external time-varying covariates, so that (4.1) holds and we may use equations (4.2) and (4.3) interchangeably to represent the mean vectors.

Settings where inference about the regression parameter β is the primary focus have been called **studies of dependence** by Cox (1972); this is in contrast to **studies of association**, where the primary focus may be the conditional expectations, $E(Y_{ij}|Y_{ik}, k < j)$ or the association parameters, such as $\text{cov}(Y_{ij}, Y_{ik})$ for $j \neq k$. In this paper we will focus on inferences about β; the variance-covariance matrix of the Y_{ij}'s is regarded as a nuisance parameter vector, although it has an important role in efficiency considerations. Individual subjects will be assumed independent throughout. We will review and contrast two general approaches for estimating β and its standard error: likelihood analysis and and generalized estimating equations (GEE), pointing out the similarities and differences in the two general approaches.

A common feature of most longitudinal studies involving human subjects is missing observations. Both the GEE and the likelihood approaches can deal with missing data in the sense that subjects with incomplete $\mathbf{Y}$ vectors may be included in the analysis. However, the resulting estimate of β and its standard error may not be consistent. The validity of inferences depends both on the method of analysis and the assumptions about the missingness mechanism.

This paper is organized as follows. In the next section we review the GEE approach. Section 4.4 reviews likelihood-based approaches in general, and two in particular, one based on the multivariate normal and another based on the multinomial. Section 4.5 discusses the validity of inferences with missing data, and section 4.6 gives a short example. Section 4.7 introduces a different class of models often used to analyse longitudinal data, mixed effects or subject-specific models, and discusses their relation to the marginal regression models. We close with some remarks about current research problems.

4.3 The GEE approach to longitudinal data analysis

The GEE approach to estimating the parameter β in the marginal regression model was introduced in companion papers by Liang and Zeger (1986) and Zeger and Liang (1986). It can be viewed as an extension of Wedderburn's (1974) quasi-likelihood method to the case with correlated responses. The approach is attractive because it requires making an assumption only about the regression model for $E(Y_{ij}|\mathbf{X}_i)$ as defined in (4.1) and (4.2); valid inferences for the regression coefficients can be obtained for the class of generalized linear marginal regression models without making distributional assumptions about the joint distribution of the Y_{ij}'s. Besides the identity or logit link, the log link is often used, especially for modelling the mean of counts. In their original papers, Liang and Zeger motivated the GEE by extending the class of generalized linear models for univariate regression analysis (McCullagh and Nelder, 1989, Chapter 2) to the multivariate setting. We will take the same approach here, first reviewing the generalized regression model for univariate responses, and noting the close connection between maximum likelihood for the exponential family and a general estimating equation (or quasi-likelihood) approach in the univariate case.

In the univariate case, the response on the ith subject, Y_i, is taken to be a scalar random variable, each subject being independently distributed with a common exponential family distribution. The mean, μ_i, of Y_i is determined by a canonical parameter θ_i, and the variance v_i, of Y_i, is determined by a dispersion parameter σ^2 and the mean μ_i via

$$\text{var}(Y_i) = v_i = \sigma^2 a(\mu_i),$$

where $a(\cdot)$ is known, σ^2 may or may not be known, and θ_i is often taken to be the natural parameter i.e. $\theta_i = \mathbf{x}_i^T\beta$, although this is not strictly necessary. For example, linear regression with homoscedastic errors corresponds to normal Y_i with mean $\mathbf{x}_i^T\beta$ (identity link) and $v_i = \sigma^2$, so that $a(\mu_i) = 1$. Logistic regression corresponds to binary Y_i, with the logit being the canonical link, $\sigma^2 = 1$ and $a(\mu_i) = \mu_i(1 - \mu_i)$. Poisson regression with the canonical link assumes each Y_i is Poisson (distributed) with $\log \mu_i = \mathbf{x}_i^T\beta, \sigma^2 = 1$ and $a(\mu_i) = \mu_i$. Other examples are gamma and inverse gamma; see Table 2.1 in McCullagh and Nelder (1989).

For this general class of exponential family models, it is easy to show, via chain rule arguments, that the ith subject's contribution to the likelihood equations for β is

$$\frac{\partial \ell_i}{\partial \beta} = \frac{\partial \mu_i}{\partial \beta}\left(\frac{\partial \theta_i}{\partial \mu_i}\frac{\partial \ell_i}{\partial \theta_i}\right) = \frac{\partial \mu_i}{\partial \beta} v_i^{-1}(y_i - \mu_i),$$

where $\ell_i = \log f(y_i|\beta, \sigma^2, \mathbf{x}_i)$ and $\partial \ell_i/\partial \beta$ is a $p \times 1$ column vector with elements $\partial \ell_i/\partial \beta_j, j = 1, \ldots, p$. The second equality follows from well-known identities for exponential families and canonical parameters. From this it follows that the likelihood equations for β are given by

$$\sum_{i=1}^{N} \left(\frac{\partial \mu_i}{\partial \beta}\right) v_i^{-1}(y_i - \mu_i) = 0. \tag{4.4}$$

Further, taking the expectation of

$$\sum_{i=1}^{N} \left(\frac{\partial \ell_i}{\partial \beta}\right)\left(\frac{\partial \ell_i}{\partial \beta}\right)^T$$

to find the Fisher information matrix, we have that the asymptotic variance-covariance matrix of β is

$$\begin{aligned} &\left\{\sum_{i=1}^{N} E\left(\frac{\partial \mu_i}{\partial \beta}\right) v_i^{-1}(Y_i - \mu_i)^2 v_i^{-1}\left(\frac{\partial \mu_i}{\partial \beta}\right)^T\right\}^{-1} = \\ &\left\{\sum_{i=1}^{N} \left(\frac{\partial \mu_i}{\partial \beta}\right) v_i^{-1}\left(\frac{\partial \mu_i}{\partial \beta}\right)^T\right\}^{-1} \end{aligned} \tag{4.5}$$

Although (4.4) is derived formally from likelihood theory, in the absence of distributional assumptions about the Y_i's, the theory of estimating equations may be used to show that, under mild regularity conditions, the solution of (4.4) still yields a consistent, asymptotically normal, estimate of β provided only that we have the regression model

$$g(\mu_i) = \mathbf{x}_i^T \beta \tag{4.6}$$

for $E(Y_i|\mathbf{x}_i)$ and independence of the Y_i's (Fahrmeir and Kaufmann, 1985; Manski, 1988; Newey and McFadden, 1993). If in addition, the model for the variance is correct, that is $\text{var}(Y_i|\mathbf{x}_i) = \sigma^2 a(\mu_i)$, then $\hat{\beta}$ is also semi-parametric efficient (Chamberlain, 1987) in the model defined solely by (4.6), in the sense of Begun, *et al.* (1983). That is, there exists no estimate of β that is locally uniformly asymptotically normal and unbiased for all distributions of the data that satisfy (4.6) with asymptotic variance less than that of $\hat{\beta}$. In this case the asymptotic variance of $\hat{\beta}$ is given by the probability limit of the right-hand side of (4.5).

Notice that σ^2 drops out of the estimating equation (4.4) for β, and all the remaining terms are functions only of the data and β. Given $\hat{\beta}$ as the solution to (4.4), a consistent estimate of σ^2 is

$$\hat{\sigma}^2 = \sum_{i=1}^{N} \frac{(Y_i - \hat{\mu}_i)^2}{N a(\hat{\mu}_i)},$$

so that $\text{var}(\hat{\beta})$ can be estimated as in equation (4.5), with $\partial\mu_i/\partial\beta$ evaluated at $\hat{\beta}$, and v_i evaluated as $\hat{\sigma}^2 a(\hat{\mu}_i)$.

If the model for v_i is not correct, $\hat{\beta}$ is still consistent but not necessarily efficient and equation (4.5) does not give a consistent estimate of var $(\hat{\beta})$. In this case we may use instead an estimate of variability proposed by Huber (1967) and White (1982),

$$\text{var}\left(\sqrt{N}\hat{\beta}\right) = \hat{\mathbf{I}}_0^{-1}\hat{\mathbf{I}}_1\hat{\mathbf{I}}_0^{-1},$$

where

$$\mathbf{I}_0 = \frac{1}{N}\left\{\sum_{i=1}^{N}\left(\frac{\partial\mu_i}{\partial\beta}\right) a(\mu_i)^{-1}\left(\frac{\partial\mu_i}{\partial\beta}\right)^T\right\},$$

$$\mathbf{I}_1 = \frac{1}{N}\left\{\sum_{i=1}^{N}\left(\frac{\partial\mu_i}{\partial\beta}\right) a(\mu_i)^{-1}(Y_i-\mu_i)^2 a(\mu_i)^{-1}\left(\frac{\partial\mu_i}{\partial\beta}\right)^T\right\},$$

and $\hat{\mathbf{I}}_0$ and $\hat{\mathbf{I}}_1$ indicate $\mathbf{I}_0$ and $\mathbf{I}_1$ with β, μ_i and $a(\mu_i)$ evaluated at $\hat{\beta}$.

Thus to summarize, for the univariate case, likelihood theory provides us with a method of obtaining estimates of the regression parameters in generalized linear regression models, which are still consistent and asymptotically normal even if some of the distributional assumptions are incorrect. Furthermore, for the estimates to be semi-parametric efficient, we need only our assumption about $v_i = \text{var}(Y_i|\mathbf{x}_i)$ to be correct. And we have a method of getting valid standard errors without assuming v_i is modelled correctly.

Liang and Zeger's approach was simply to extend equation (4.4) in the obvious way to accommodate the generalized multivariate regression model defined by equations (4.1)–(4.3). In the multivariate case, we have vectors $\mathbf{Y}_i$ and $\mu_i = E(\mathbf{Y}_i|\mathbf{X}_i)$, and $\text{var}(\mathbf{Y}_i) = \mathbf{\Sigma}_i$ for some appropriately specified positive definite $T \times T$ matrix $\mathbf{\Sigma}_i$. If we know $\mathbf{\Sigma}_i$, the obvious extension of (4.4) is to estimate β by solving

$$\sum_{i=1}^{N} \left(\frac{\partial \mu_i}{\partial \beta}\right)^T \mathbf{\Sigma}_i^{-1}(\mathbf{y}_i - \mu_i) = \mathbf{0}, \tag{4.7}$$

where now $(\partial \mu_i/\partial \beta)^T$ denotes a $p \times T$ matrix whose (j,k)th element is $\partial \mu_{ik}/\partial \beta_j$. Still assuming $\mathbf{\Sigma}_i$ known, the ideas in Fahrmeir and Kaufmann (1985), Manski (1988) and Newey and McFadden (1993) can be used to show that under mild regularity conditions, the solution to (4.7) is an asymptotically normal, efficient and consistent estimate of β, and its asymptotic variance is given by

$$\text{var}(\hat{\beta}) = \left\{\sum_{i=1}^{N} \left(\frac{\partial \mu_i}{\partial \beta}\right)^T \mathbf{\Sigma}_i^{-1} \frac{\partial \mu_i}{\partial \beta}\right\}^{-1}. \tag{4.8}$$

Two questions come immediately to mind. Is there a class of multivariate exponential family likelihoods for $f(\mathbf{Y}_i|\mathbf{X}_i, \beta, \mathbf{\Sigma}_i)$ that gives rise to likelihood equations similar to (4.7)? What happens when $\mathbf{\Sigma}_i$ is unknown, or estimated by $\mathbf{V}_i$, and the estimate is used in (4.7) in place of $\mathbf{\Sigma}_i$? The first question was addressed by McCullagh and Nelder (1989, Chapter 9) using a quasi-likelihood approach, and will be taken up in the next section. With regard to the second issue, Liang and Zeger (1986) show that if $\mathbf{\Sigma}_i$ is replaced by some positive definite matrix $\mathbf{V}_i$ (called the 'working variance matrix'), the resulting $\hat{\beta}$ remains consistent and asymptotically normal, and its asymptotic variance can be consistently estimated by

$$\text{var}\left(\sqrt{N}\hat{\beta}\right) = \hat{\mathbf{I}}_0^{-1}\hat{\mathbf{I}}_1\hat{\mathbf{I}}_0^{-1}, \tag{4.9}$$

where

$$\hat{\mathbf{I}}_0 = \frac{1}{N}\left\{\sum_{i=1}^{N} \left(\frac{\partial \hat{\mu}_i}{\partial \beta}\right)^T \mathbf{V}_i^{-1} \frac{\partial \hat{\mu}_i}{\partial \beta}\right\}$$

and

$$\hat{\mathbf{I}}_1 = \frac{1}{N}\left\{\sum_{i=1}^{N}\left(\frac{\partial\hat{\mu}_i}{\partial\beta}\right)^T \mathbf{V}_i^{-1}(\mathbf{Y}_i - \hat{\mu}_i)(\mathbf{Y}_i - \hat{\mu}_i)^T\mathbf{V}_i^{-1}\frac{\partial\hat{\mu}_i}{\partial\beta}\right\}.$$

Furthermore, if $\mathbf{V}_i$ consistently estimates $\text{var}(\mathbf{Y}_i) = \Sigma_i$ at a rate of at least $N^{1/4+\delta}, \delta > 0$, then the resulting estimate of β is also semiparametric efficient in the model defined solely by equation (4.2), and an estimate of its variance is given by $\hat{\mathbf{I}}_0$.

In summary, one can get a consistent estimate for β and its standard error without making any assumptions about the distribution of $\mathbf{Y}_i$ other than its mean, but for efficiency we need to specify Σ_i and find a consistent estimate of it. One could, for example, simply set $\mathbf{V}_i = \mathbf{I}$, but, depending upon the design, this might lead to a quite inefficient estimate of β. This point will be explored further in section 4.8.

So the question arises, how shall we specify and estimate $\text{var}(\mathbf{Y}_i|\mathbf{X}_i)$? This is considerably more difficult than in the univariate case for several reasons. First of all, we need to specify correlations which, unlike the variance, may bear no obvious functional relationship to the mean. Second, even for the variance, the straightforward univariate extension of setting

$$\text{var}(Y_{ij}|\mathbf{X}_i) = \sigma^2 a(\mu_{ij}) \tag{4.10}$$

is not always so attractive, since it implies that, apart from its dependence on μ_{ij}, the variance is constant over time. If we assume, for example, the normal variance function, $a(\mu_{ij}) = 1$, this implies constant variance over time which is often not realistic for longitudinal studies. For the binary case, the variance function $a(\mu_{ij}) = \mu_{ij}(1 - \mu_{ij})$ clearly still holds.

The approach of Liang and Zeger (1986; also Zeger and Liang, 1986) was to use the univariate approach, i.e. assume (4.10) holds, and specify a 'working correlation matrix,' $\mathbf{R}_i$, which may be easier to model and estimate. For example, setting $\mathbf{R}_i = \mathbf{I}$ yields the 'independence' working assumption. The assumption that for all $j \neq k$

$$\text{corr}(Y_{ij}, Y_{ik}) = \rho$$

yields the 'exchangeable' working assumption, and

$$\mathbf{R}_i = \begin{bmatrix} 1 & \rho & \cdots & \rho \\ \vdots & & & \vdots \\ \rho & \rho & \cdots & 1 \end{bmatrix}.$$

Letting $\mathbf{A}_i$ be a diagonal matrix with $a(\mu_{ij})$ on the diagonal, we then specify the working variance as

$$\mathbf{V}_i = \sigma^2 \mathbf{A}_i^{1/2} \mathbf{R}_i \mathbf{A}_i^{1/2}. \tag{4.11}$$

We can now estimate β in two stages. First set $\mathbf{R}_i = \mathbf{I}$, so that the estimating equations

$$\sum_{i=1}^{N} \left(\frac{\partial \mu_i}{\partial \beta}\right)^T \mathbf{A}_i^{-1}(\mathbf{y}_i - \mu_i) = \mathbf{0} \tag{4.12}$$

are only a function of the data $(\mathbf{Y}_i, \mathbf{X}_i)$ and the unknown β. Solve (4.12) to get a provisional estimate of β and thus μ_i, say $\tilde{\mu}_i$, and use this to estimate $\mathbf{R}_i$ and σ^2 from the residuals

$$\mathbf{r}_i = \tilde{\mathbf{A}}^{-1/2}(\mathbf{y}_i - \tilde{\mu}_i)(\mathbf{y}_i - \tilde{\mu}_i)\tilde{\mathbf{A}}_i^{-1/2}.$$

For example, if $\mathbf{R}_i$ is exchangeable, we can estimate ρ by

$$\hat{\rho} = \frac{2}{NT(T-1)} \sum_{i=1}^{N} \sum_{j<\ell} r_{ij} r_{i\ell} \tag{4.13}$$

and

$$\hat{\sigma}^2 = \frac{1}{NT} \sum_{i=1}^{N} \sum_{j=1}^{T} r_{ij}^2. \tag{4.14}$$

Now re-estimate β via

$$\textstyle\sum_{i=1}^{N} \left(\frac{\partial \mu_i}{\partial \beta}\right)^T \mathbf{V}_i^{-1}(\mathbf{Y}_i - \mu_i) = \mathbf{0}, \tag{4.15}$$

where $\mathbf{V}_i$ is specified by (4.11), using $\hat{\mathbf{R}}_i$ for $\mathbf{R}_i$. Other choices for $\mathbf{R}_i$ and corresponding estimates are discussed in Liang and Zeger (1986) and Zeger and Liang (1986).

Clearly this approach could be generalized to model $\mathbf{V}_i$ directly rather than assuming (4.10) and modelling $\mathbf{R}_i$, in which case at stage 2, $\mathbf{V}_i$ is estimated directly from

$$\mathbf{r}_i = (\mathbf{Y}_i - \tilde{\mu}_i)(\mathbf{Y}_i - \tilde{\mu}_i)^T.$$

For example, in the case where we assume an arbitrary $T \times T$ variance-covariance matrix which does not depend upon i, we would estimate each $\mathbf{V}_i$ by

$$\hat{\mathbf{V}} = \frac{1}{N} \sum_{i=1}^{N} \mathbf{r}_i \mathbf{r}_i^T.$$

We now turn to a discussion of likelihood-based approaches.

4.4 Likelihood-based approaches to longitudinal data analysis

In this section we shall consider some likelihood-based approaches for the analysis of longitudinal data, keeping in mind that our objective is inferences about the regression parameters β in the marginal mean model for $\mu_i = E(\mathbf{Y}_i|\mathbf{X}_i)$. We continue to assume a $T \times 1$ vector is observed for each subject and independence of subjects; we discuss missing data in section 4.5.

McCullagh and Nelder (1989, Chapter 9) proposed extensions of quasi-likelihood for the general multivariate case, by considering integrals of the estimating equations for β. An alternative approach was considered by Zhao, Prentice and Self (1992) who proposed a class of likelihoods which they termed **partly exponential models** (PEMs). The PEM specifies the distribution of $\mathbf{Y}_i$ as

$$p(\mathbf{y}_i) = \Delta_i^{-1} \exp\{\theta_i^T \mathbf{y}_i + C(\mathbf{y}_i, \lambda)\}, \tag{4.16}$$

where Δ_i^{-1} is a normalizing constant which is a function of θ_i and λ, $C(\mathbf{y}_i, \lambda) = C_i(\lambda)$ is a shape function with parameter vector λ, and θ_i is a canonical parameter uniquely determined by

$$\mu_i = \int \Delta_i^{-1} \mathbf{y} e^{\{\theta_i^T \mathbf{y}_i + C_i(\lambda)\}} d\nu(\mathbf{y}), \tag{4.17}$$

where ν denotes Lebesgue or counting measure, as appropriate.

This general representation for the likelihood is similar to the exponential dispersion models discussed in Jorgensen (1987). PEMs are more general in that they allow λ to be a vector, rather than a scalar as in exponential dispersion models. This permits a broader class of association structures for the multivariate responses. It is easily shown that the PEM has many properties similar to ordinary exponential families. In particular, letting $\log p(\mathbf{y}_i) = \ell_i$ and $\mathbf{w}_i = \partial C_i(\lambda)/\partial\lambda$, we have

$$\frac{\partial \ell_i}{\partial \theta_i} = \mathbf{y}_i - \mu_i, \quad \frac{\partial \ell_i}{\partial \lambda} = \mathbf{w}_i - \eta_i \tag{4.18}$$

where $E(\mathbf{W}_i) = \eta_i$, and

$$\frac{\partial^2 \ell_i}{\partial \theta_i \partial \theta_i^T} = -\text{var}(\mathbf{Y}_i), \quad \frac{\partial^2 \ell_i}{\partial \theta_i \partial \lambda^T} = -\text{cov}(\mathbf{Y}_i, \mathbf{W}_i),$$

$$\frac{\partial^2 \ell_i}{\partial \lambda \partial \lambda^T} = -\text{var}(\mathbf{W}_i) + E\left\{\frac{\partial^2 C_i(\lambda)}{\partial \lambda \partial \lambda^T}\right\}.$$

From these it follows that

$$\frac{\partial \mu_i}{\partial \theta_i} = \text{var}(\mathbf{Y}_i), \quad \frac{\partial \mu_i}{\partial \lambda} = \text{cov}(\mathbf{Y}_i, \mathbf{W}_i). \tag{4.19}$$

These properties make it easy to take derivatives of the likelihood function. Note that in general $\mathbf{w}_i$ may depend upon both $\mathbf{y}_i$ and λ, although we usually suppress dependence on λ. In the cases we consider here, $C_i(\lambda)$ can be represented as a linear function in λ, and $\mathbf{w}_i$ does not depend upon λ, but we retain the more general definition. There are two cases of this general model which are especially useful for longitudinal data.

Case 1: Multivariate normal. In the case where $\mathbf{Y}_i$ lies in R^T, a frequently used model is to specify $\mathbf{Y}_i$ as $N(\mu_i, \Sigma)$ where Σ is an arbitrary $T \times T$ positive definite matrix. In the more general case we may want to let Σ depend upon covariates, or to be a function of some smaller set of parameters, say $\Sigma = \Sigma(\alpha)$. The standard link function for the normal is the identity, so that $\mu_i = \mathbf{X}_i\beta$, although any differentiable link function is possible. It is easily seen that we may take $\theta_i = \Sigma \mathbf{X}_i\beta$, $C_i(\lambda) = \text{tr}\Sigma^{-1}\mathbf{y}_i\mathbf{y}_i^T$, λ to be the non-redundant elements of Σ^{-1}, and $\mathbf{w}_i$ to be the non-redundant elements of $\mathbf{y}_i\mathbf{y}_i^T$. We may also take λ or Σ to depend upon a parameter vector α. The PEM thus leads to the usual multivariate normal representation for this case.

Case 2: Multinomial. Suppose each y_{ij} is dichotomous; in this case each person falls into one cell of a 2^T array which cross-classifies the T variables. Any log-linear model (Cox 1972; Bishop, Fienberg and Holland, 1975) for the cell probabilities of this array can be written as

$$p(\mathbf{y}_i) \propto \exp\left(\mathbf{y}_i^T\theta_i + \Sigma\lambda_k w_k\right),$$

where the w_{ik}'s are cross products of the y_{ij}'s, e.g.

$$y_{i1}y_{i2}, \ldots, y_{i1}y_{i2}y_{i3}, \ldots$$

and the λ_k's are the 'association terms,' e.g. the logarithms of the conditional odds ratios and linear combinations thereof. The exact specification of λ and $\mathbf{w}_i$ depends upon the choice of log-linear model. For example, for the pairwise model $\mathbf{w}_i$ is all possible pairs, $y_{i1}y_{i2} \ldots, y_{iT-1}y_{iT}$, and λ is the set of all possible conditional log-odds ratios

$$\log\left\{\frac{P(y_{i\ell}=1, y_{ik}=1|y_{ij})P(y_{i\ell}=0, y_{ik}=0|y_{ij})}{P(y_{i\ell}=1, y_{ik}=0|y_{ij})P(y_{i\ell}=0, y_{ik}=1|y_{ij})}\right\}, j \neq k, \ell.$$

This differs from the standard log-linear model in that θ_i is not allowed to vary freely, but is determined by the relations in (4.1)–(4.2) and (4.17), where we use the logit transformation $\ell(\mu_{ij}) = \text{logit}(\mu_{ij})$. We could also let λ depend upon covariates, or on a vector of smaller dimension, e.g. for the pairwise model we may take all the log-odds ratios equal to a scalar α. The use of this model for longitudinal data analysis is discussed in Fitzmaurice and Laird (1993).

Given an independent sample of size N from $p(\mathbf{y}_i)$ as specified in (4.16), we may estimate β and λ (or α) by maximum likelihood. As noted in Zhao, Prentice and Self (1992), the derivatives of the log-likelihood can be found using the chain rule and the expressions in (4.18)–(4.19) to give

$$\sum_{i=1}^{N} \begin{pmatrix} \mathbf{D}_i^T & \mathbf{0} \\ \mathbf{0} & \mathbf{F}^T \end{pmatrix} \begin{pmatrix} \boldsymbol{\Sigma}_i^{-1} & \mathbf{0} \\ \mathbf{B}_i\boldsymbol{\Sigma}_i^{-1} & \mathbf{I} \end{pmatrix} \begin{pmatrix} \mathbf{y}_i - \mu_i \\ \mathbf{w}_i - \eta_i \end{pmatrix}, \qquad (4.20)$$

where $\mathbf{D}_i = \partial\mu_i/\partial\beta_i$, $\mathbf{F} = \partial\lambda/\partial\alpha$, $\boldsymbol{\Sigma}_i = \text{var}(\mathbf{Y}_i)$ and $\mathbf{B}_i = \text{cov}(\mathbf{Y}_i, \mathbf{W}_i)$.

Setting the derivatives to zero gives for $\hat{\beta}$ the estimating equation

$$\sum_{i=1}^{N} \left(\frac{\partial\mu_i}{\partial\beta}\right)^T \hat{\boldsymbol{\Sigma}}_i^{-1} (\mathbf{y}_i - \hat{\mu}_i) = \mathbf{0}, \qquad (4.21)$$

and for $\hat{\lambda}$

$$\mathbf{F}^T \left\{ \sum_{i=1}^{N} \hat{\mathbf{B}}_i\hat{\boldsymbol{\Sigma}}_i^{-1} (\mathbf{y}_i - \hat{\mu}_i) + \sum_{i=1}^{N} (\mathbf{w}_i - \hat{\eta}_i) \right\} = \mathbf{0}. \qquad (4.22)$$

Also, as noted in Zhao, Prentice and Self (1992), the asymptotic variance matrix of $(\hat{\beta}, \hat{\alpha})$ is

$$\text{var}(\hat{\beta}, \hat{\alpha}) =$$

$$\begin{bmatrix} \sum_{i=1}^{N} \mathbf{D}_i^T\boldsymbol{\Sigma}_i^{-1}\mathbf{D}_i & \mathbf{0} \\ \mathbf{0} & \sum_{i=1}^{N} \mathbf{F}^T \left\{\text{var}(\mathbf{W}_i) - \mathbf{B}_i\boldsymbol{\Sigma}_i^{-1}\mathbf{B}_i^T\right\} \mathbf{F} \end{bmatrix}^{-1}. \qquad (4.23)$$

Notice that the estimating equation for $\hat{\beta}$ is identical to the GEE given in (4.15), apart from the specification of $\boldsymbol{\Sigma}_i$ rather than $\mathbf{V}_i$. Since $\mathbf{V}_i$ can be arbitrarily specified, the ML estimator with complete data falls into the class of GEE estimators; thus they have the same properties. Zhao, Prentice and Self (1992) recommend using the 'robust' variance estimate (4.9) for $\hat{\beta}$ rather than (4.23) above, to guard against the possibility that the likelihood, and hence var$(\mathbf{Y}_i)$, is misspecified.

The estimate of β is consistent even if the likelihood is misspecified, because the expected value of the score equations for β is zero provided only that μ_i is correctly specified. This follows because the score equations for β depend on the data only through $\mathbf{y}_i - \mu_i$. Other parametric representations for $f(\mathbf{Y}_i|\mathbf{X}_i)$, e.g. Bahadur's (1961) representation for each y_{ij} discrete, may give ML estimates which have smaller asymptotic variance than the solution to (4.21), but those estimates will not generally be consistent for β if their assumed distribution fails to hold.

In general, Fisher scoring can be used to compute $(\hat{\beta}, \hat{\alpha})$, although, as Zhao, Prentice and Self (1992) note, it is complicated to implement whenever the quantities $\mathbf{V}_i$, $\mathbf{B}_i$ and var$(\mathbf{W}_i)$ are not simple functions of $\hat{\beta}$ and $\hat{\lambda}$.

For the multivariate normal case, with linear link function $\mu_i = \mathbf{X}_i\beta$ and common variance $\mathbf{\Sigma}(\alpha)$, it is straightforward to show that the likelihood equations for β and α respectively can be written as

$$\sum_{i=1}^{N} \mathbf{X}_i^T \hat{\mathbf{\Sigma}}^{-1} (\mathbf{y}_i - \mathbf{X}_i\hat{\beta}) = \mathbf{0}$$

$$\sum_{i=1}^{N} \hat{\mathbf{F}}^T \hat{\mathbf{\Phi}}^{-1} (\mathbf{w}_i - \hat{\sigma}) = \mathbf{0},$$

where $\hat{\sigma}$ and $\mathbf{w}_i$ are the non-redundant elements of $\mathbf{\Sigma}$ and $(\mathbf{y}_i - \mu)(\mathbf{y}_i - \mu_i)^T$, respectively, and $\mathbf{\Phi} = $ var$(\mathbf{W}_i)$; see Zhao and Prentice (1990). When $\mathbf{\Sigma}$ is completely unstructured, so that $\mathbf{F}^T = \mathbf{I}$, these equations can be readily solved by iteratively reweighted least squares:

$$\begin{aligned} \hat{\beta} &= (\textstyle\sum_{i=1}^{N} \mathbf{X}_i^T \hat{\mathbf{\Sigma}}^{-1} \mathbf{X}_i)^{-1} \textstyle\sum_{i=1}^{N} \mathbf{X}_i^T \hat{\mathbf{\Sigma}}^{-1} \mathbf{y}_i, \\ \hat{\mathbf{\Sigma}} &= \tfrac{1}{N} \textstyle\sum_{i=1}^{N} (\mathbf{y}_i - \mathbf{X}_i\hat{\beta})(\mathbf{y}_i - \mathbf{X}_i\hat{\beta})^T. \end{aligned}$$

For the multinomial case where each y_{ij} is dichotomous, it is natural to use the logit link for μ_{ij}, hence $\partial\mu_i/\partial\beta = \mathbf{\Gamma}_i\mathbf{X}_i$, where $\mathbf{\Gamma}_i = \text{diag}\{\mu_{ij}(1-\mu_{ij}),\ j = 1, \ldots, T\}$. The likelihood equations thus become

$$\textstyle\sum_{i=1}^{N} \mathbf{X}_i^T \hat{\mathbf{\Gamma}}_i \hat{\mathbf{\Sigma}}_i^{-1} (\mathbf{y}_i - \hat{\mu}_i) = \mathbf{0}$$

and

$$\mathbf{F}^T \left\{ \textstyle\sum_{i=1}^{N} \hat{\mathbf{B}}_i \hat{\mathbf{V}}_i^{-1} (\mathbf{y}_i - \hat{\mu}_i) + \textstyle\sum_{i=1}^{N} (\mathbf{w}_i - \hat{\eta}_i) \right\} = \mathbf{0},$$

where $\mathbf{\Sigma}_i = $ var$(\mathbf{Y}_i)$ and $\mathbf{W}_i$ is some set of cross products in the Y_{ij}'s, depending upon the log-linear model selected to specify the association.

Although $\mathbf{\Sigma}_i, \mathbf{B}_i$ and η_i are straightforward to calculate if we know the joint probabilities $p(\mathbf{y}_i)$, there is no general expression for the cell probabilities as a function of (β, α) for an arbitrary log-linear model. However, iterative proportional fitting can easily be used to circumvent this problem; see Fitzmaurice and Laird (1993) for details.

4.5 Estimation with incomplete data

In many settings, some responses may be missing on some individuals. In this case we may write $\mathbf{y}_i^T = (\mathbf{y}_{io}, \mathbf{y}_{im})$, where $\mathbf{y}_{io}$ is the $r_i \times 1$ vector of responses obtained for the ith person and $\mathbf{y}_{im}$ is the $m_i \times 1$ vector of responses which are not observed, where $T = r_i + m_i$. We can similarly partition

$$\mu_i^T = (\mu_{io}^T, \mu_{im}^T), \ \mathbf{X}_i^T = (\mathbf{X}_{io}^T, \mathbf{X}_{im}^T),$$

and $\mathbf{V}_i$ and Σ_i in the form

$$\mathbf{V}_i = \begin{bmatrix} \mathbf{V}_{io} & \mathbf{V}_{iom} \\ \mathbf{V}_{imo} & \mathbf{V}_{im} \end{bmatrix}, \quad \Sigma_i = \begin{bmatrix} \Sigma_{io} & \Sigma_{iom} \\ \Sigma_{imo} & \Sigma_{im} \end{bmatrix}.$$

The GEE approach to handling missing data is to still use equation (4.15) to estimate β, but to replace $\mathbf{y}_i$, μ_i, $\mathbf{V}_i$ and $\partial\mu_i/\partial\beta$ by $\mathbf{y}_{io}$, μ_{io}, $\mathbf{V}_{io}$ and $\partial\mu_{io}/\partial\beta$, yielding

$$\sum_{i=1}^N \left(\frac{\partial\mu_{io}}{\partial\beta}\right)^T \hat{\mathbf{V}}_{io}^{-1} (\mathbf{y}_{io} - \hat{\mu}_{io}) = 0. \qquad (4.24)$$

This still gives a consistent estimate of $\hat{\beta}$, provided $E(\mathbf{Y}_{io}) = \mu_{io}$. Implicitly we condition on $\mathbf{y}_{io}$ being observed; hence $E(\mathbf{Y}_{io})$ in general depends upon the missingness process. Little and Rubin (1987) give a convenient typology for characterizing the missingness process by defining $p(\mathbf{R}_i)$, where $\mathbf{R}_i$ is a $T \times 1$ response indicator vector with $R_{ij} = 1$ if y_{ij} is observed and $R_{ij} = 0$ otherwise. If $p(\mathbf{R}_i)$ does not depend upon $\mathbf{y}_{io}$ or $\mathbf{y}_{im}$ then the process is said to be **missing completely at random** (MCAR). It is easily seen that under MCAR, $E(\mathbf{Y}_{io}) = E(\mathbf{Y}_{io}|\mathbf{R}_i) = \mu_{io}$, so that consistency of the GEE estimator holds. If $p(\mathbf{R}_i)$ does not depend on $\mathbf{y}_{im}$ or (β, α) but may depend upon $\mathbf{y}_{io}$, then the missingness process is said to be **missing at random** (MAR), and ignorable. In this case, $E(\mathbf{Y}_{io}|\mathbf{R}_i)$ is not in general equal to μ_{io} and consistency of the GEE may not hold. However, the contribution of $p(\mathbf{R}_i)$ to the likelihood may be ignored, and we find the maximum likelihood estimator by maximizing

$$\mathcal{L}(\beta, \lambda) = \sum \log p_{io}(\mathbf{y}_{io}),$$

where

$$p_{io}(\mathbf{y}_{io}) = \int p(\mathbf{y}_i) d\nu(\mathbf{y}_{im})$$

is the ordinary marginal density of $\mathbf{y}_{io}$. Although one may test the MCAR assumption against a specific MAR alternative, it is not in general possible to test whether or not $p(\mathbf{R}_i)$ depends upon $\mathbf{Y}_{im}$.

Using the form of $p(\mathbf{y}_i)$ in equation (4.16) and taking derivatives under the integral sign, we can easily derive a set of equations for the derivatives of $\log p_{io}$ parallel to equation (4.18) which are similar to those given in Dempster, Laird and Rubin (1977) for incomplete data from exponential families:

$$\frac{\partial \log p_{oi}}{\partial \theta_i} = E(\mathbf{Y}_i|\mathbf{y}_{io}) - \mu_i, \quad \frac{\partial \log p_{oi}}{\partial \lambda} = E(\mathbf{W}_i|\mathbf{y}_{io}) - \eta_i.$$

Writing $E(\mathbf{Y}_i|\mathbf{y}_{io})$ as $\mathcal{E}(\mathbf{Y}_i)$ and $E(\mathbf{W}_i|\mathbf{y}_i)$ as $\mathcal{E}(\mathbf{W}_i)$, we have that the likelihood equations for (β, α) can be written as

$$\sum_{i=1}^{N} \begin{bmatrix} \left(\frac{\partial \mu_i}{\partial \beta}\right)^T & 0 \\ 0 & \mathbf{F}^T \end{bmatrix} \begin{bmatrix} \hat{\Sigma}_i^{-1} & 0 \\ \hat{\mathbf{B}}_i\hat{\Sigma}_i^{-1} & \mathbf{I} \end{bmatrix} \begin{pmatrix} \mathcal{E}(\mathbf{Y}_i) - \hat{\mu}_i \\ \mathcal{E}(\mathbf{W}_i) - \hat{\nu}_i \end{pmatrix} = 0.$$

The equation for $\hat{\beta}$ becomes

$$\sum_{i=1}^{N} \left(\frac{\partial \mu_i}{\partial \beta}\right)^T \hat{\Sigma}_i^{-1} \{\mathcal{E}(\mathbf{Y}_i) - \hat{\mu}_i\} = 0, \tag{4.25}$$

which is identical to (4.21), except that $\mathbf{y}_i$ is replaced by $\mathcal{E}(\mathbf{Y}_i)$.

In general, equation (4.25) is not equivalent to equation (4.24) which defines the GEE with missing data, except when $E(\mathbf{Y}_i|\mathbf{y}_{io})$ is linear in $\mathbf{y}_{io}$, or when $\mathbf{y}_{im}$ and $\mathbf{y}_{io}$ are assumed independent. To see this, note that if $\mathcal{E}(\mathbf{y}_i)$ is linear in $\mathbf{y}_{io}$ we may write

$$\mathcal{E}(\mathbf{Y}_i) = \begin{pmatrix} \mathbf{y}_{io} \\ \mu_{io} + \Sigma_{iom}\Sigma_{io}^{-1}(\mathbf{y}_{io} - \mu_{io}) \end{pmatrix},$$

so that (4.25) can be written

$$\sum_{i=1}^{N} \left(\frac{\partial \mu_i}{\partial \beta}\right)^T \hat{\Sigma}_i^{-1} \begin{bmatrix} \mathbf{I} \\ \hat{\Sigma}_{imo}\hat{\Sigma}_{io}^{-1} \end{bmatrix} (\mathbf{y}_{io} - \hat{\mu}_{io}) = 0.$$

However, using standard identities for the inverse of a partitioned matrix, we can show that

$$\Sigma_i^{-1} \begin{bmatrix} \mathbf{I} \\ \Sigma_{imo}\Sigma_{io}^{-1} \end{bmatrix} = \begin{bmatrix} \Sigma_{io}^{-1} \\ 0 \end{bmatrix},$$

so equation (4.25) is now identical to equation (4.24). Likewise, if $\mathbf{y}_{im}$ is independent of $\mathbf{y}_{io}$,

$$\mathcal{E}(\mathbf{Y}_i) - \mu_i = \begin{pmatrix} \mathbf{y}_{io} - \mu_{io} \\ 0 \end{pmatrix}$$

and $\mathbf{\Sigma}_{imo} = \mathbf{0}$, so that equation (4.25) again reduces to equation (4.24).

Thus in two special cases, linearity of $\mathcal{E}(\mathbf{Y}_i)$ in $\mathbf{y}_{io}$, and independence of $\mathbf{y}_{io}$ and $\mathbf{y}_{im}$, the forms of the ML and GEE estimators of β coincide with missing data, apart from the fact that the variance-covariance matrix will generally be specified and estimated differently. Thus in these two cases, the ML estimator will be consistent if the missingness is MCAR even if the likelihood is misspecified. Notice that this does not imply consistency of an arbitrary GEE under MAR if linearity or independence holds: for this, correct specification of the variance-covariance matrix is needed. Another bonus of these two assumptions (linearity or independence) is that the equations for $\hat{\beta}$ do not depend upon $\mathbf{X}_{im}$, although in general the equations for $\hat{\alpha}$ will.

The asymptotic variance of the ML estimators under the PEM $(\hat{\beta}, \hat{\alpha})$ can be approximated by the inverse of the observed information. In general, the second derivatives of the log-likelihood are quite complex, so we use instead the inverse of the square of the observed scores:

$$\text{var}\left(\hat{\beta}, \hat{\alpha}\right) = \left[\sum_{1}^{N} \begin{pmatrix} \frac{\partial \ell_{oi}}{\partial \beta} \\ \frac{\partial \ell_{oi}}{\partial \alpha} \end{pmatrix} \begin{pmatrix} \frac{\partial \ell_{oi}}{\partial \beta} \\ \frac{\partial \ell_{oi}}{\partial \alpha} \end{pmatrix}^{T}\right]^{-1}. \quad (4.26)$$

The form of the likelihood equations naturally suggests the EM algorithm for computation. At the pth E-step we compute

$$\mathbf{y}_i^{(p)} = E\left(\mathbf{Y}_i | \mathbf{y}_{io}, \hat{\beta}^{(p)}, \hat{\alpha}^{(p)}\right), \mathbf{w}_i^{(p)} = E\left(\mathbf{W}_i | \mathbf{y}_{io}, \hat{\beta}^{(p)}, \hat{\alpha}^{(p)}\right),$$

and then at the M-step solve the complete data equations for $\hat{\beta}^{(p+1)}$ and $\hat{\alpha}^{(p+1)}$, treating $\mathbf{y}_i^{(p)}$ and $\mathbf{w}_i^{(p)}$ as $\mathbf{y}_i$ and $\mathbf{w}_i$. In practice, we may simply take one iteration of Fisher scoring at the M-step rather than iterating to convergence.

This connection with the EM algorithm also shows the relationship to 'imputation' methods, which are based on imputing values for $\mathbf{y}_{im}$, and then doing the usual complete data computations. In general, ML will differ from imputation methods even if imputation is based on $\mathcal{E}(\mathbf{Y}_{im})$, because (i) iterative imputation is required unless the missingness follows a monotone pattern, and (ii) substituting $\hat{\mathbf{y}}_{im} = \mathcal{E}(\mathbf{Y}_{im})$ for the missing $\mathbf{y}_{im}$ into $\partial C_i(\lambda)/\partial\lambda$ in order to compute $\mathbf{w}_i$ does not yield $\mathcal{E}(\mathbf{W}_i)$. However, in practice, the solution for $\hat{\beta}$ under ML may be similar to that given by imputation if the imputed values are based on the model specified by $\mathcal{E}(\mathbf{Y}_{im})$. Returning to the multivariate normal case, because $E(\mathbf{Y}_i|\mathbf{y}_{io})$ is linear in $\mathbf{y}_{io}$, the likelihood equations for $\hat{\beta}$ again reduce to iteratively reweighted least squares, involving only the 'observed' y's

$$\hat{\beta} = \sum_{i=1}^{N} \left(\mathbf{X}_{io}^T \hat{\Sigma}_{io}^{-1} \mathbf{X}_{io} \right)^{-1} \sum_{i=1}^{N} \mathbf{X}_{io}^T \hat{\Sigma}_{io}^{-1} \mathbf{y}_{io}.$$

For Σ unstructured, we must solve

$$\hat{\Sigma} = \frac{1}{N} \sum_{i=1}^{N} \mathcal{E}\left(\mathbf{W}_i\right)$$

where

$$\mathbf{W}_i = \left(\mathbf{y}_i - \mathbf{X}_i\hat{\beta}\right)\left(\mathbf{y}_i - \mathbf{X}_i\hat{\beta}\right)^T.$$

A simple expression for this can be given by writing $\mathbf{y}_{io} = \mathbf{T}_i\mathbf{y}_i$, where $\mathbf{T}_i$ is a $T_i \times T$ matrix obtained from a $T \times T$ identity matrix by removing the rows corresponding to the missing observations. Then it is straightforward to show that

$$\begin{aligned} \mathcal{E}(\mathbf{W}_i) &= \Sigma\mathbf{T}_i^T\Sigma_{io}^{-1}(\mathbf{y}_{io} - \mu_{io})(\mathbf{y}_{io} - \mu_{io})^T\Sigma_{io}^{-1}\mathbf{T}_i\Sigma \\ &+ \Sigma - \Sigma\mathbf{T}_i^T\Sigma_{io}^{-1}\mathbf{T}_i\Sigma \\ &= \mathbf{U}_i, \end{aligned}$$

say, so that the likelihood equations for Σ can be expressed as

$$\hat{\Sigma} = \frac{1}{N}\sum_{i=1}^{N}\hat{\mathbf{U}}_i.$$

Maximum likelihood estimation for the multivariate normal regression model is now commercially available through BMDP-5V or SAS `PROC MIXED`. Both of these programs permit arbitrary specification for the design matrix, $\mathbf{X}_i$, and allow arbitrary patterns of missing data, but assume the identity link. Both packages also enable the user to specify a variety of models for the variance-covariance matrix, Σ, for example random effects, autoregressive, banded, etc. In addition, restricted maximum likelihood estimates of the variance parameters may be obtained as an alternative to maximum likelihood.

For the multinomial case, it is straightforward to calculate both $E(\mathbf{Y}_i|\mathbf{y}_{io})$ and $E(\mathbf{W}_i|\mathbf{y}_{io})$ once we obtain the underlying cell probabilities for the 2^T table, because the basic distribution is multinomial. This we must do at each iteration using iterative proportional fitting in the complete data case in order to compute β_i, Σ_i, etc., hence the additional computing burden is not excessive. Notice that for this case $\mathcal{E}(\mathbf{Y}_i)$ is not linear in $\mathbf{y}_{io}$ unless $T = 2$, so that the equations for $\hat{\beta}$ do not reduce to the GEE equations. For this case, we need to specify $\mathbf{X}_{im}$ to compute the ML estimates.

In summary, the GEE method is quite easy to extend when not all subjects are measured at every occasion by utilizing what can be called the 'all available observations' method. Although it is easily implemented, it does not, in general, yield consistent estimates of the regression vector β, except under the most benign assumption (MCAR) on the missingness process. The likelihood approach can be generalized in a straightforward way to accommodate missing responses under a more general assumption (MAR) on the missingness process. Its implementation can be carried out using the EM algorithm, although this may be complicated in practice since the M-step is iterative. The fact that the likelihood method requires milder assumptions on the missingness process than does the GEE must be balanced against the fact that the likelihood method now requires the correct distributional specification for validity of the estimation method. In the case where we assume multivariate normal, the GEE and likelihood-based methods give similar estimates for β with missing data, although the estimates of the variance-covariance parameters can be very different (Park, 1993). Some limited simulation results show that for the case with discrete outcomes, the likelihood assumption is robust to misspecification in that the ML estimator has only small bias under MAR when the likelihood is misspecified (Fitzmaurice, Laird and Rotnitzky, 1993). Additional work in this area to study the sensitivity of the likelihood specification under MAR would be desirable.

4.6 Example: childhood obesity

The Muscatine coronary risk factor study was a longitudinal survey of school-age children carried out in Muscatine, Iowa (Woolson and Clarke, 1984). Each child was measured biannually from 1971 to 1981 and classified as obese or not obese. There is considerable incomplete data on children who did not participate in all the survey years. Our objective here is to characterize the sex differences in trends in obesity. For simplicity, we consider only the cohort ages 7–9 in 1977, who were measured again in 1979 and 1981. This included 1014 children; only 460 of these were measured on all three surveys. The data are shown in Tables 4.1 and 4.2.

The marginal expectation of response, which in this case represents the probability μ_{ij} that a given child is obese at a particular age, is modelled as a logistic function of age (linear and quadratic terms), gender and their interaction. Two parallel analyses are carried out for these data, one using the GEE, and one using a likelihood-based approach. For the GEE, we consider two differing correlation assumptions: the independence and the exchangeable correlation. Estimated regression

Table 4.1 *Muscatine study: all data on males. Note: 1 = obese; 0 = not obese; * = missing*

	Child's Obesity Status			
	Age 8	Age 10	Age 12	Count
None Missing	1	1	1	20
	1	1	0	7
	1	0	1	9
	1	0	0	8
	0	1	1	8
	0	1	0	8
	0	0	1	15
	0	0	0	15
Missing Time 1	*	1	1	13
	*	1	0	3
	*	0	1	2
	*	0	0	42
Missing Time 2	1	*	1	3
	1	*	0	1
	0	*	1	6
	0	*	0	16
Missing Time 3	1	1	*	11
	1	0	*	1
	0	1	*	3
	0	0	*	38
Missing Times 1,2	*	*	1	14
	*	*	0	55
Missing Times 1,3	*	1	*	4
	*	0	*	33
Missing Times 2,3	1	*	*	7
	0	*	*	45

coefficients, their standard errors calculated using (4.9) and Z-statistics are given in Table 4.3. Results suggest a strong increase in prevalence of obesity as children age. Girls appear to have a higher initial increase and lower subsequent decrease than boys, but these results are not statistically significant. For the likelihood-based analysis we assumed a saturated association structure, i.e., we fit a model with all two-way interactions, and a three-way interaction term. The parameter estimates and their standard errors, calculated from (4.26), are given in Table 4.4. All of the association parameters differ significantly from

Table 4.2 *Muscatine study: all data on females. Note: 1 = obese; 0 = not obese; * = missing*

	Child's Obesity Status			
	Age 8	Age 10	Age 12	Count
None Missing	1	1	1	21
	1	1	0	6
	1	0	1	6
	1	0	0	2
	0	1	1	19
	0	1	0	13
	0	0	1	14
	0	0	0	154
Missing Time 1	*	1	1	8
	*	1	0	1
	*	0	1	4
	*	0	0	47
Missing Time 2	1	*	1	4
	1	*	0	0
	0	*	0	16
	0	*	1	3
Missing Time 3	1	1	*	11
	1	0	*	1
	0	1	*	3
	0	0	*	25
Missing Times 1,2	*	*	1	13
	*	*	0	39
Missing Times 1,3	*	1	*	5
	*	0	*	23
Missing Times 2,3	1	*	*	7
	0	*	*	47

zero. Despite this, and the large fraction of missing observations, there is remarkably good agreement between the likelihood-based results and the GEE using the exchangeable correlation assumption. There are more pronounced differences between the two versions of the GEE, one using the independence assumption and one using the exchangeable correlation. We might have expected the independence assumption to perform poorly here because of the high degree of missing data.

Table 4.3 *Estimated regression parameters* and standard errors for the data presented in Tables 4.1 and 4.2: GEE analysis*

	Independence		Exchangeable	
Covariate	Est ± SE	Z	Est ± SE	Z
INT	-1.330 ± 0.990	-13.36	-1.250 ± 0.097	-13.86
SEX	0.034 ± 0.142	0.24	0.040 ± 0.139	0.29
AGE(L)	0.146 ± 0.074	1.96	0.146 ± 0.069	2.11
AGE(Q)	0.012 ± 0.038	0.33	0.015 ± 0.036	0.43
SEX*AGE(L)	0.113 ± 0.106	107	0.158 ± 0.099	1.60
SEX*AGE(Q)	-0.082 ± 0.053	-1.56	-0.089 ± 0.050	-1.78
CORRELATION			$\hat{\rho} = 0.448$	

*See text for a description of the regression model.

4.7 Random effects models

A seemingly different approach to the analysis of longitudinal data has its origins in the growth curve literature, where investigators were concerned with studying individual patterns of growth, and how patterns of growth might differ among subsets of the population. The general approach taken in this setting can be characterized as a two-stage random effects approach. In stage 1, a growth model is assumed to hold for each subject separately, with subject-specific parameters. At stage 2, these subject-specific parameters are treated as random variables which are modelled as functions of subject characteristics or other covariates.

In general, this two-stage approach leads to a class of models and methods that differ from the marginal regression models which we discussed in the preceding sections, with a few important exceptions. With linear growth curves and linear link functions, a typical stage 1 individual regression model assumes

$$E(\mathbf{Y}_i|\beta_i) = \mathbf{Z}_i\beta_i, \text{var}(\mathbf{Y}_i|\beta_i) = \sigma^2\mathbf{I},$$

where $\mathbf{Z}_i$ specifies the design on time for the ith subject, and β_i is the coefficient vector for the ith subject. For example, if each subject changes linearly over time, $\mathbf{Z}_i$ will be an $R_i \times 2$ matrix with a column of ones, and a column of times of observations. At stage 2, treating the subject effects as random, we posit a multivariate regression model of the form

$$\beta_i = \mathbf{A}_i\beta + \mathbf{b}_i,$$

where $\mathbf{A}_i$ are fixed subject covariates (e.g., sex, race), β are the mean parameters, and $\mathbf{b}_i$ is the subject-level error term, independent of $\mathbf{e}_i$, with mean zero and var($\mathbf{b}_i$) $= \mathbf{D}$, say. Here $\mathbf{b}_i$ has dimension equal to the

Table 4.4 *Estimated regression parameters* and standard errors for the data presented in Tables 4.1 and 4.2: likelihood analysis*

			Association Parameters		
Covariate	Est ± SE	*Z*	Parameter	Est ± SE	Z
INT	−1.360 ± 0.980	−13.85	α_{12}	3.31 ± 0.48	6.95
SEX	0.043 ± 0.138	0.31	α_{13}	2.76 ± 0.45	6.15
AGE(L)	0.142 ± 0.063	2.27	α_{23}	2.82 ± 0.34	8.32
AGE(Q)	0.014 ± 0.038	0.40	α_{123}	−1.81 ± 0.62	−2.90
SEX*AGE(L)	0.162 ± 0.096	1.68			
SEX*AGE(Q)	−0.089 ± 0.049	−1.81			

*See text for a description of the regression model.

parameter vector in stage 1. Stage 1 and stage 2 together imply that

$$E(\mathbf{Y}_i) = \mathbf{Z}_i\{\mathbf{A}_i\beta + E(\mathbf{b}_i)\} = \mathbf{Z}_i\mathbf{A}_i\beta$$

and

$$\text{var}(\mathbf{Y}_i) = E\{\text{var}\,(\mathbf{Y}_i|\beta_i)\} + \text{var}\,\{E(\mathbf{Y}_i|\beta_i)\} = \sigma^2\mathbf{I} + \mathbf{Z}_i\mathbf{D}\mathbf{Z}_i^T.$$

Thus for this case, we again have a marginal regression model for $E(\mathbf{Y}_i)$, with a special structure for the design matrix, $\mathbf{X}_i = \mathbf{Z}_i\mathbf{A}_i$, and for the variance covariance matrix, $\Sigma_i = \text{var}(\mathbf{Y}_i)$, which depends upon the design on time. Note that one attractive feature of modelling the data in this manner is that each subject can have a different number of observations, as well as a different design matrix, $\mathbf{Z}_i$.

In the case where the error terms $\mathbf{e}_i$ and $\mathbf{b}_i$ are assumed to be normally distributed, this model is just a special case of the general multivariate normal regression model, and parameter estimates of β, σ^2 and $\mathbf{D}$ and their standard errors can be obtained using SAS Proc Mixed or BMDP-5V.

One feature of this approach that sets it apart from the general marginal regression model is the presence of the subject-specific random effects $\mathbf{b}_i$ or $\beta_i = \mathbf{A}_i\beta + \mathbf{b}_i$. These random effects can be estimated, using parametric empirical Bayes (EB) or best linear unbiased predictors (BLUPs) as

$$\hat{\mathbf{b}}_i = \hat{\mathbf{D}}\mathbf{Z}_i^T\Sigma_i^{-1}(\mathbf{y}_i - \mathbf{X}_i\hat{\beta})$$

or

$$\hat{\beta}_i = \mathbf{A}_i\hat{\beta} + \hat{\mathbf{b}}_i.$$

These estimators are termed empirical Bayes because they can be expressed as the posterior mean of β_i given the data $\mathbf{Y}_i$. They represent a compromise between using the overall model-based mean $\mathbf{A}_i\hat{\beta}$, and the individual least-squares regression estimators, $(\mathbf{Z}_i^T\mathbf{Z}_i)^{-1}\mathbf{Z}_i^T\mathbf{y}_i$.

Estimates of individual growth curves are useful in a number of settings: to identify outlying individuals, to check model goodness of fit, or to make individual predictions. These EB estimators are also useful in a different context, where our ultimate objective is to predict individual survival, or other important endpoints, as a function of repeatedly measured subject outcomes. This arises, for example, in developing surrogate markers for disease progression, such as CD4 counts as a marker for the onset of AIDS. What the EB estimates allow one to do is construct individual profiles at each point in time, regardless of the pattern of measurements available, so that current values, or any feature of past history, are available to use as predictors of patient failure. An example of this, involving the use of CD4 counts to predict survival in

AIDS patients, is given in Tsiatis, DeGruttola and Wulfsohn (1995).

We remark that these subject-specific models are quite popular in the nonlinear case, where at stage 1 the response can be modelled as some nonlinear function of time and subject-specific parameters, i.e.

$$y_{ij} = g(\mathbf{z}_{ij}, \beta_i) + e_{ij}$$

where the e_{ij}'s are independent $N(0, \sigma^2)$ measurement error terms and $\mathbf{z}_{ij}$ gives the design for the ith subject at the jth time. This does not fall in the class of generalized linear regression models, but the maximum likelihood theory for normally distributed error terms and subject-specific effects can be extended straightforwardly. The computational issues are more complex because there is in general no closed-form solution for the likelihood function or its derivatives (Lindstrom and Bates, 1990).

When we do not use the identity link, then the two-stage approach generally does not lead to the same class of marginal regression models discussed in the previous sections. Consider a two-stage model of the form:

Stage 1:

$$E(Y_{ij}|\beta_i, \mathbf{Z}_i) = \mu_{ij}, g(\mu_{ij}) = \mathbf{z}_{ij}^T\beta_i,$$
$$\text{var}(Y_{ij}|\beta_i) = \sigma^2 a(\mu_{ij}), \text{cov}(Y_{ij}, Y_{ik}|\beta_i) = 0 \text{ for } (j \neq k).$$

and

Stage 2:

$$\beta_i = \mathbf{A}_i\beta + \mathbf{b}_i,$$

where $E(\mathbf{b}_i) = \mathbf{0}, \text{var}(\mathbf{b}_i) = \mathbf{D}$. Such models have been proposed by several authors, including Stiratelli, Laird and Ware (1984), Wong and Mason (1985) and Gilmour, Anderson and Rae (1985). A broad class of completely specified distributions can be obtained by assuming independence of the Y_{ij}'s given β_i at stage 1, with the Y_{ij}'s following a specified exponential family model where $g(\cdot)$ is the canonical link, and assuming multivariate normality at stage 2 for the $\mathbf{b}_i$.

From stage 1 and 2, it follows that

$$E\{g(\mu_{ij})\} = \mathbf{A}_i\beta,$$

but this differs from our marginal regression model which requires $g\{E(Y_{ij})\}$ to be a linear function of β and covariates. Thus, although this class of models is useful in the context of individual curve fitting, its use in settings where the primary interest is inference about marginal regression parameters is more controversial; see Zeger, Liang and Albert (1988), Laird (1991) and Neuhaus, Kalbfleisch and Hauck (1991) for further discussion. The problem stems from the fact the β is the regression parameter in the regression of $\mathbf{Y}_i$ on both $\mathbf{b}_i$ and $\mathbf{Z}_i$. But conditioning on the $\mathbf{b}_i$'s is quite different from conditioning on known, observable subject or design characteristics, since the $\mathbf{b}_i$'s represent inherently unobservable subject characteristics. Hence components of β cannot be interpreted in the way that we usually interpret regression coefficients, i.e. the effect of changing one variable while holding the rest fixed, the problem being that what we mean by 'the rest' is ambiguous when $\mathbf{b}_i$ is part of the variable set we condition on.

Consider, for example, a cross-over trial designed to compare two treatments A and B for pain relief of a transient condition. Assume the outcome in each period is determined by a binary response, and that, ignoring period and carry-over effects, we model the logit of p(relief) as linear in treatment and subject effects:

$$\text{logit}\, \mu_{ij} = \beta_0 + \beta_1 T_{ij} + b_i,$$

Here T_{ij} is 1 if treatment A is assigned to subject i in period j, and 0 otherwise, and b_i measures the propensity of a subject to have pain relief, regardless of treatment. A large treatment effect in a marginal regression model would imply that subjects were much more likely to have pain relief under treatment A than under treatment B. If most patients either always or never experience pain relief under any treatment, then the marginal regression parameter for treatment will inevitably be small. In the conditional model, however, a large value of β_1 can result even if treatment A is only effective in a small subset of patients who experience different results under different drugs, since we condition on a patient's propensity to experience pain relief. In this case, where the two regression models disagree, it may still be meaningful to estimate either effect, but one must be mindful that the estimated parameter needs to be interpreted appropriately.

As pointed out in Diggle, Liang and Zeger (1994), one special case where the nonlinear random effects approach leads to a marginal model occurs when we use the log-link and a single random intercept, i.e.

$$\log \mu_{ij} = \mathbf{Z}_{ij}^T \beta + b_{0i},$$

where b_{0i} is a zero-mean, subject-specific random effect. In this case

$$\log E(Y_{ij}) = \beta_0^* + \mathbf{Z}_{ij}\beta,$$

where $\beta_0^* = \log E\left(e^{b_{0i}}\right)$. Thus apart from the intercept term, the β's are identical to the marginal regression parameters.

4.8 Discussion

An overview of a broad area of methodology such as the analysis of panel data must necessarily give limited attention to many important topics. This review has focused on the fitting of marginal regression models because of their prominent role in applications. They are useful in implementing multivariate analyses in a variety of settings, including longitudinal studies of growth and ageing as well as experimental trials. In addition, these marginal regression models can be used to analyse multivariate data, clustered data and repeated measures data.

We have discussed three general analytical approaches for estimation of the regression parameters and their standard errors: a generalized estimating equations approach which can be viewed as the extension of the generalized linear model to handle correlated data; a maximum likelihood approach based on either the multivariate normal data for measured responses, or the multinomial model for dichotomous outcomes; and two-stage random effects models. In the case of the general linear model with continuous outcomes, these three approaches to model development and estimation lead to very similar analyses in large samples. Although the GEE and likelihood-based approaches generally do not agree for binary data, in practice they can give reasonably similar results, even with a substantial amount of missing data, provided some attention is given to specifying the correlation structure.

The specification of the association structure in the data plays an important role in determining the efficiency of the estimates of β, as well as in determining bias when data are missing at random. Although the most efficient estimator requires specification, and estimation of $\text{var}(\mathbf{Y}_i)$, in practice $\text{var}(\mathbf{Y}_i)$ may be quite misspecified in some settings with little reduction in efficiency. This issue has been studied by many authors, including Bloomfield and Watson (1975), Zhao, Prentice and Self (1992), Fitzmaurice, Laird and Rotnitzky (1993), and Lipsitz *et al.* (1994).

The loss of efficiency due to misspecification of $\text{var}(\mathbf{Y}_i)$ is minimal in simple balanced designs where each subject has the same design on time. This will happen in a typical panel study where all subjects are measured at the same set of fixed time points, and all other covariates are fixed over time. In this case, each subject contributes roughly the same amount of information about the parameter estimates, hence little efficiency is lost by sub-optimal weighting. The presence of imbalance due to missing responses on some subjects, or due to designs with time-varying covariates which differ across individuals, can lead to considerable loss of inefficiency when $\text{var}(\mathbf{Y}_i)$ is misspecified, since subjects will not be weighted optimally. Examples include some higher-order cross-over designs for estimating carry-over effects (Fitzmaurice, Laird and Rotnitsky, 1993), and designs with stochastically varying covariates (Lipsitz *et al.*, 1994).

In closing, we briefly review one area of much current research: handling drop-outs. Panel studies dealing with prolonged follow-up on human subjects are inevitably subject to attrition. From one point of view, this may be regarded simply as a special case of missing data, where the missingness follows a monotone pattern. However, there is very often reason to believe that subjects who discontinue participation differ systematically from those who continue, so that attrition may introduce selection effects into a study if not properly accounted for. This is especially true in studies of disease, and in clinical trials, where subjects who are doing poorly, or in some cases exceptionally well, may drop out, leaving one with a biased sample at the end of follow-up. The term **informative drop-out** is sometimes used to describe such situations.

In principle, informative drop-outs can be handled simply as missing data in likelihood models, provided that the probability of drop-out depends only upon variables observed prior to drop-out, since in this case drop-out can be viewed as a special case of MAR data. In practice, however, for unbiasedness to hold, all the variables predicting the probability of drop-out must be included in the likelihood model for the outcomes, even if the association between these variables and the outcomes is not of direct interest. Thus in practice, model specification may be quite complex. As an example, subjects may drop out of a clinical trial because of the occurrence of side effects; yet intention-to-treat inferences about the effects of the drug on outcome are not conditioned on presence or absence of side effects. Recent work on GEE methods with MAR data proposes to estimate the probability of drop-out separately, then use inverse probability weighting of observed cases to correct for selection effects (Robins, Rotnitzky and Zhao, 1995).

A second difficulty arises with the MAR assumption itself. MAR implies that the distribution of responses at time $t + 1$, conditional on the past history, is the same for all subjects regardless of drop-out status at time t. If the assumption fails, the process is said to be **non-ignorable** (Little and Rubin, 1987), since valid inferences in this setting generally require specification of a model for the probability of drop-out and simultaneous estimation of the parameters of both the non-response and the data model. Such models are difficult to specify and estimate, since for non-ignorable non-response, the probability of drop-out depends upon unobserved response or characteristics.

Current work on modelling the non-response mechanism falls into two general categories. In the longitudinal data setting, one approach, called the **selection model approach**, follows the work of Heckman (1976), who proposed methods for adjusting for selection bias in univariate regression models. For monotone missing patterns, Diggle and Kenward (1994) proposed to model the probability of drop-out at time t as a function of the unobserved response at t, and also possibly previously observed responses. As in the usual selection models, the estimated regression parameters β can be quite sensitive to model specification. Wu and Carroll (1988) proposed a different approach based on the two-stage random effects model whereby the probability of drop-out at time t depends upon the unobserved subject-specific random effects, e.g. slope. They assumed that drop-outs could only occur at a set of discrete measurement times. Subsequent work on this model has relaxed that assumption. Schluchter (1992) and DeGruttola and Tu (1995) assume that the time to drop-out, or some suitable transformation, is normal, again with mean depending upon the unobserved random subject effects. Recent work (Tsiatis, DeGruttola and Wulfsohn, 1995) has relaxed the normality assumption, fitting a proportional hazards regression model for the the probability of drop-out at time t, and estimating the underlying hazard nonparametrically.

The second general approach taken to model informative drop-outs is to 'reverse the conditioning' and posit a marginal model for drop-out (not conditioned on the outcomes Y_{ij} or the random effects) and the conditional distribution of outcome given time of drop-out. This approach is sometimes called a **mixture model approach**, since inferences about outcome unconditional on drop-out status are obtained by mixing over the distribution of drop-out time. Subjects who complete the study are treated as a separate conditioning stratum. Wu and Bailey (1988; 1989) developed an early version of the mixture model as an approximation to the selection model analysis because of the computational complexity involved in fitting the selection model. They again used the random effects approach and assumed that the random slope depends upon drop-out time. They developed approximate methods for estimation based on fitting analysis of covariance type models to a subjects slope, estimated by ordinary least squares. Mori, Woodworth and Woolson (1992) use the same model, but estimate the mean slope using the average empirical Bayes estimates of the individual slopes, conditioned on drop-out time. When using the random effects model, the mixture model is easily implemented using standard software, since the general random effects model discussed in section 4.7 allows one to specify the random effects (the β_i of section 4.7) as dependent upon any number of regressors.

Analytic methods for handling longitudinal data with missing observations, time-varying covariates, and generalized linear regression models for a variety of outcomes have progressed rapidly in the last 10–15 years, and we are beginning to see a first generation of commercial software packages emerge for fitting. Work on informative drop-outs is more recent, and there is much current activity. We can look forward to having better analytic techniques for handling informative drop-outs in the near future.

Acknowledgements

The author is indebted to Sir David Cox and Dr Andrea Rotnitzky for their help in the preparation of this manuscript. This work was supported by grant GM 27948 from the National Institute of Health.

References

Bahadur, R.R. (1961) A representation of the joint distribution of responses to n dichotomous items. In H. Solomon (ed.), *Studies in Item Analysis and Prediction*, Stanford Mathematical Studies in the Social Sciences VI. Stanford University Press, Stanford, CA.

Begun, J.M., Hall, W.J., Huang, W.M. and Wellner, J.A. (1983) Information and asymptotic efficiency in parametric-nonparametric models. *Annals of Statistics* **11**, 432–452.

Bishop, Y.M.M., Fienberg, S.E. and Holland, P.W. (1975) *Discrete Multivariate Analysis: Theory and Practice*. MIT Press, Cambridge, MA.

Bloomfield, P. and Watson, G.S. (1975) The inefficiency of least squares. *Biometrika*, **62**, 121–128.

Chamberlain, G. (1987) Asymptotic efficiency in estimation with conditional moment restrictions. *Journal of Econometrics* **34**, 305–342.

Cox, D.R. (1972) The analysis of multivariate binary data. *Applied Statistics*, **21**, 113–120.

DeGruttola, V. and Tu, X.M. (1995) Modeling progression of CD-4 lymphocyte count and its relation to survival time. *Biometrics*, **50**, 1003-1014.

Dempster, A.P., Laird, N.M. and Rubin, D.B. (1977) Maximum likelihood estimation from incomplete data via the EM algorithm (with discussion). *Journal of the Royal Statistical Society, Series B*, **39**, 1–38.

Diggle, P.J. and Kenward, M.G. (1994) Informative drop-out in longitudinal data analyses (with discussion). *Applied Statistics*, **43**, 49–93.

Diggle, P.J., Liang, K.-Y. and Zeger, S.L. (1994) *Analysis of Longitudinal Data*. Clarendon Press, Oxford.

Fahrmeir, L. and Kaufmann, H. (1985) Consistency and asymptotic normality of the maximum likelihood estimator in generalized linear models. *Annals of Statistics*, **13**, 342–368.

Fitzmaurice, G.M. and Laird, N.M. (1993) A likelihood-based method for analyzing longitudinal binary responses. *Biometrika*, **80**, 141–151.

Fitzmaurice, G.M., Laird, N.M. and Rotnitzky, A.G. (1993) Regression models for discrete longitudinal responses. *Statistical Science*, **8**, 284–309.

Gilmour, A.R., Anderson, R.D. and Rae, A.L. (1985) The analysis of binomial data by a generalized linear mixed model. *Biometrika*, **72**, 593–599.

Heckman, J.J. (1976) The common structure of statistical models of truncation, sample selection, and limited dependent variables and a simple estimator for such models. *Annals Economic Social Measurement*, **5**, 475–492.

Huber, P. (1967) The behavior of maximum likelihood estimators under nonstandard conditions. In L.M. LeCam and J. Neyman (eds), *Proceedings of the Fifth Berkeley Symposium on Mathematical Statistics and Probability*, **1**, University of California Press, Berkeley, pp. 221–233.

Jorgensen, B. (1987) Exponential dispersion models (with discussion). *Journal of the Royal Statistical Society, Series B*, **49**, 127–162.

Laird, N.M. (1991) Topics in likelihood-based methods for longitudinal data analysis. *Statistica Sinica*, **1**, 35–50.

Liang, K.Y. and Zeger, S.L. (1986) Longitudinal data analysis using generalized linear models. *Biometrika*, **73**, 13–22.

Lindstrom, M.J. and Bates, D.M. (1990) Nonlinear mixed effects models for repeated measures data. *Biometrics*, **46**, 673–689.

Lipsitz, S.R., Fitzmaurice, G.M., Orav, E.J. and Laird, N.M. (1994) Performance of generalized estimating equations in practical situations. *Biometrics*, **50**, 270–278.

Little, R.J.A., and Rubin, D.B. (1987) *Statistical Analysis with Missing Data*. John Wiley, New York.

Manski, C.F. (1988) *Analog Estimation in Econometrics*. Chapman & Hall, New York.

McCullagh, P. and Nelder, J.A. (1989) *Generalized Linear Models* 2nd edn Chapman & Hall, New York.

Mori, M., Woodworth, G.G. and Woolson, R.F. (1992) Application of empirical Bayes inference to estimation of rate of change in the presence of informative right censoring. *Statistics in Medicine*, **11**, 621–631.

Neuhaus, J.M., Kalbfleisch, J.D. and Hauck, W.W. (1991) A comparison of cluster-specific and population-averaged approaches for analyzing correlated binary data. *International Statistical Review*, **59**, 25–35.

Newey, W.K. and McFadden, D. (1993) Estimation in large samples. In D. McFadden and R. Engle (eds) *Handbook of Econometrics*, Vol 4. North-Holland, Amsterdam.

Park, T. (1993). A comparison of the generalized estimating equation approach with the maximum likelihood approach for repeated measurements. *Statistics in Medicine*, **12**, 1723–1732.

Robins, J.M., Rotnitzky, A.G., and Zhao, L.P. (1995) Analysis of semi-parametric regression models for repeated outcomes in the presence of missing data. *Journal of the American Statistical Association*, **90**, 106–121.

Schluchter, M.D. (1992) Methods for the analysis of informatively censored longitudinal data. *Statistics in Medicine*, **11**, 1861–1870.

Stiratelli, R., Laird, N. and Ware, J.H. (1984) Random-effects models for serial observations with binary response. *Biometrics*, **40**, 961–971.

Tsiatis, A., DeGruttola, V. and Wulfsohn, M. (1995) Modeling the relationship of survival to longitudinal data measured with error; applications to patients in AIDS. *Journal of the American Statistical Association*, **90**, 27–37.

Waternaux, C., Laird, N.M. and Ware, J.H. (1989) Methods for analysis of longitudinal data: blood-lead concentrations and cognitive development. *Journal of the American Statistical Association*, **84**, 33–41.

Wedderburn, R.W.M. (1974) Quasilikelihood functions, generalized linear models and the Gauss–Newton method. *Biometrika*, **61**, 439–447.

White, H. (1982) Maximum likelihood estimation of misspecified models. *Econometrics*, **50**, 1–25.

Wong, G.Y. and Mason, W.M. (1985) The hierarchical logistic regression model for multilevel analysis. *Journal of the American Statistical Association*, **80**, 513–525.

Woolson, R.F. and Clarke, W.R. (1984) Analysis of categorical incomplete longitudinal data. *Journal of the Royal Statistical Society, Series A*, **147**, 87–99.

Wu, M.C. and Bailey, K.R. (1988) Analyzing changes in the presence of informative right censoring caused by death and withdrawal. *Statistics in Medicine*, **7**, 337–346.

Wu, M.C. and Bailey, K.R. (1989) Estimation and comparison of changes in the presence of informative right censoring: conditional linear model. *Biometrics*, **45**, 939–955.

Wu, M.C. and Carroll, R.J. (1988) Estimation and comparison of changes in the presence of informative right censoring by modeling the censoring process. *Biometrics*, **44**, 175–188.

Zeger, S.L. and Liang, K.Y. (1986) Longitudinal data analysis for discrete and continuous outcomes. *Biometrics*, **42**, 121–130.

Zeger, S.L., Liang, K.Y., and Albert, P. (1988) Models for longitudinal data, a generalized estimating equation approach. *Biometrics*, **44**, 1049–1060.

Zhao, L.P. and Prentice, R.L. (1990) Correlated binary regression using a quadratic exponential model. *Biometrika*, **77**, 642–648.

Zhao, L.P., Prentice, R.L., and Self, S.G. (1992) Multivariate mean parameter estimation by using a partly exponential model. *Journal of the Royal Statistical Society, Series B*, **54**, 805–811.

CHAPTER 5

Pricing by no arbitrage

Bjarne Astrup Jensen and Jørgen Aase Nielsen

5.1 Introduction

The pricing of financial assets has long been a focal topic for practitioners in the financial markets as well as for researchers in economics and finance. However, for the last 15 years the topic has also appealed to researchers with a background in mathematics with special emphasis on probability theory. Simultaneously, the pricing models developed and their degree of sophistication have triggered a rapidly growing statistical literature developing statistical techniques for parameter estimation and process identification.

Theories and models dealing with price formation in financial markets are divided into (at least) two markedly different types. One type of model attempts to explain levels of asset prices, risk premiums etc., in an *absolute* manner in terms of so-called 'fundamentals'. Well-known models of this type include the so-called 'rational expectation model' equating stock price with the discounted value of expected future dividends. Another type of model has a more modest scope, namely to explain in a *relative* manner some asset prices in terms of other given and observable asset prices. It is this latter kind of pricing theory we deal with in this paper.

Price formation in financial markets has several characteristics that render the behaviour of these markets as close to the long-standing and analytically convenient textbook paradigm of perfect divisibility and instantaneous spread of information as can be found anywhere in the real world. Quantity discounts are rare events in financial markets, and indivisibility problems are usually ignorable. Consequently, price formation is characterized by **linearity**.

The dual entrance to the pricing of financial assets is caused by these facts. The economics involved in relative pricing of financial assets leads to linear pricing functionals applied to the random payments from financial assets, whereas probability theory and modern martingale theory provide theorems of mathematical representations of such linear functionals. Loosely speaking, there is an equivalence between the economic phenomenon of assets being priced by certain linear functionals, and the representation of asset prices as martingales under certain probability measures.

This equivalence arises from the very simple concept of **arbitrage** in economic theory. In earlier economic literature the concept of 'arbitrage' had a slightly different meaning from that generally associated with the word today. An arbitrageur was a banker involved in finding the most profitable way of executing a payment order on behalf of a customer, usually a payment order involving the exchange of one currency for another in relation to import/export activities. The basic assumption simply states that all participants in the market prefer more to less – they are 'greedy' – and that any increase in consumption opportunities must somehow be paid for. The latter assumption is the famous 'no free lunch' condition.

These assumptions are very weak indeed. Apart from being 'greedy' – meaning that anyone able to obtain anything of value for free will not hesitate to do so – market participants can be heterogeneous with respect to their preferences for consumption over time and states of the economy: some might be risk-averse and others might have an attraction towards risk, some might have outspoken preferences for immediate consumption and others might have preferences for saving in order to ensure their future consumption opportunities. The only common characteristic needed is that they should be greedy.

The concept of 'no arbitrage' is so simple that it seems to have been ignored in the academic world because of the wrong perception that such a trivial concept could not lead to any new insight. A notable exception is the famous Modigliani–Miller argument (Modigliani and Miller, 1958), for the irrelevance of the capital structure decision, although their argument is qualitatively different from the later development in financial economics.

However, the rapid development in financial economics that took place from the early 1970s onwards has proved that arguments based on 'no arbitrage' can lead quite far. Since the seminal paper by Black and Scholes (1973) – and other contemporaneous, but less well-known, contributions – the methodology based on 'no arbitrage' has been the cornerstone of pricing theories of derivatives and so-called 'financial engineering'.

Although we do not restrict the preferences, it is clear that the mere *raison d'être* of financial markets has to do with differences in preferences and the demand from market participants to redistribute (i) consumption opportunities over time and (ii) risk across individuals. The allocation resulting from this redistribution reflects exactly the best possible fulfilment of the consumers' preferences. This means that the prices of the already existing assets in the financial markets must be expected to reflect these two motives.

Financial assets can be thought of as specific mixtures of some fundamental building blocks representing the risk components and the sequence of dates at which payments can take place. If we are able to extract the prices of these fundamental building blocks from the prices of bonds, shares and other financial assets traded in the market, we can create and price – through the linearity of pricing – new assets simply by choosing new mixtures of the building blocks. In some cases all fundamental building blocks can be priced uniquely. Hence, new assets can be priced uniquely. This market situation is denoted as a **complete** market. In other situations only a subset of prices of fundamental building blocks can be extracted. This market situation is denoted as an **incomplete** market. In this latter case, pricing bounds are the best that can be obtained for new assets without introducing investor-specific preferences in the analysis.

The outline of this paper is as follows. In section 5.2 some basic mathematical notation is defined. In sections 5.3 to 5.6 the most basic model, involving only one future date, is described, and the economic concepts are given a precise mathematical definition within this very basic framework.

Section 5.3 contains the fundamental, although rather elementary, Theorem 1 characterizing the equivalence between 'no arbitrage' and linear pricing. Sections 5.4 and 5.5 introduce the concepts of equivalent martingale measures as corollaries to Theorem 1. Section 5.6 introduces the equivalence between linear pricing and the martingale representation of prices. Throughout, numerical examples are given in order to illustrate the theory.

Sections 5.7 to 5.9 deal with multiperiod models in discrete time. Despite the fact that this is mostly a modulation of the basic theme from sections 5.3 to 5.6 and a derivation of statements parallel to those found there, the framework is much richer and calls for the introduction of a number of new concepts with precise mathematical definitions. A number of numerical examples are given, and we discuss the concepts of 'splitting index' and 'martingale basis' as a measure of the intensity of information arrival.

Sections 5.10 to 5.13 deal with continuous-time models. While the space at our disposal enables us to give a self-contained tutorial presentation for the discrete-time models, this is not possible for the continuous-time models. The technicalities involved are quite subtle. However, little new economic insight is gained compared to the discrete-time models. We have therefore tried to present, according to our own judgement, the most important parts of the theory and the parallels to the statements about discrete-time models without giving full proofs of every detail. Section 5.14 contains concluding remarks and supplies references to a number of sources that the reader can consult in order to get such full and rigorous proofs and where the topics covered can be pursued in further detail.

5.2 Notation

The notation in $\mathcal{R}^m$ is defined as follows:

- By $\mathcal{R}_+^m$ we denote the set of vectors $\boldsymbol{x} \in \mathcal{R}^m \backslash \{\mathbf{0}\}$, with $x_j \geq 0$, $j = 1, 2, \ldots, m$.
- By $\mathcal{R}_{++}^m$ we denote the set of vectors $\boldsymbol{x} \in \mathcal{R}^m$ with $x_j > 0$, $j = 1, 2, \ldots, m$.
- By the vector inequality $\boldsymbol{x} < \boldsymbol{y}$ $(\boldsymbol{x} > \boldsymbol{y})$ for $\boldsymbol{x}, \boldsymbol{y} \in \mathcal{R}^m$ we mean $\boldsymbol{y} - \boldsymbol{x} \in \mathcal{R}_+^m$ $(\boldsymbol{x} - \boldsymbol{y} \in \mathcal{R}_+^m)$.
- By the vector inequality $\boldsymbol{x} << \boldsymbol{y}$ $(\boldsymbol{x} >> \boldsymbol{y})$ we mean $\boldsymbol{x} - \boldsymbol{y} \in \mathcal{R}_{++}^m$ $(\boldsymbol{y} - \boldsymbol{x} \in \mathcal{R}_{++}^m)$.
- Both $\boldsymbol{x} \cdot \boldsymbol{y}$ and $<\boldsymbol{x}, \boldsymbol{y}>$, $\boldsymbol{x}, \boldsymbol{y} \in \mathcal{R}^m$, are expressions for the Euclidean inner product $\sum_{i=1}^m x_i y_i$. If $\boldsymbol{A}$ is a matrix we denote analogously $\boldsymbol{A} \cdot \boldsymbol{x}$ and $\boldsymbol{x} \cdot \boldsymbol{A}$ as the usual vector-matrix multiplication.
- By $\boldsymbol{xy}$, $\boldsymbol{x}, \boldsymbol{y} \in \mathcal{R}^m$, we mean the function $j \to x_j y_j$ defined on the set $\{1, 2, \ldots, m\}$. If $\boldsymbol{A}$ is a $n \times m$ matrix, we define analogously $\boldsymbol{Ax}$ as the function $j \to (\boldsymbol{A} \cdot \boldsymbol{x})_j$.
- Any probability measure P on $\{1, 2, \ldots, m\}$ defines an inner product, denoted by $< \boldsymbol{x}, \boldsymbol{y} >_P$, on $\mathcal{R}^m$:

$$< \boldsymbol{x}, \boldsymbol{y} >_P \equiv E^P[\boldsymbol{xy}] = \sum_{\omega=1}^m P_\omega x_\omega y_\omega . \tag{5.1}$$

When $P_\omega > 0$ for all $\omega = 1, 2, \ldots, m$, the topology induced by $<,>_P$ is equivalent to the usual Euclidean topology. ∎

Let there be given a **probability space** $(\Omega, \mathcal{F}, P)$ and a **filtration** $\{\mathcal{F}_\tau\}_{\tau \in \mathcal{T}}$. In terms of this notation, $\mathcal{T}$ denotes the set of time indices, where trading in financial assets is allowed or where it is possible either to withdraw money from the portfolio in order to consume or to invest additional money into the portfolio.

The simplest possible scenario is a two-period model, i.e.

$$\mathcal{T} = \{0, T\},$$

and the simplest possible scenario with respect to Ω is the finite state space equipped with the discrete σ-algebra:

$$\Omega = 1, 2, \ldots, S;\ \mathcal{F} = 2^{\Omega};\ \mathcal{F}_0 = \{\emptyset, \Omega\};$$
$$\mathcal{F}_T = \mathcal{F};\ P_\omega > 0\ \ \forall \omega \in \Omega.$$

The financial market has $N + 1$ traded assets, whose prices at time 0 are denoted by the vector $\boldsymbol{s}_0 \in \mathcal{R}^{N+1}$, $\boldsymbol{s}_0 = \left(s_0^0, s_0^1, \ldots, s_0^N\right)$. At time T, the owner of financial asset number n receives a random payment $\boldsymbol{s}_T^n = (s_T^n(1), \ldots, s_T^n(\omega), \ldots, s_T^n(S))$. The matrix symbol $\boldsymbol{S}$ denotes the $(N + 1) \times S$ matrix whose rows are the vectors $\boldsymbol{s}_T^n$. The column vectors of $\boldsymbol{S}$ are denoted by $\boldsymbol{s}_T(\omega)$.

At time $t = 0$ the agents can buy and sell financial assets. The portfolio position for an individual agent is given in terms of the vector $\boldsymbol{\theta} \in \mathcal{R}^{N+1}$, where θ_n denotes the quantity of the nth asset bought at time 0. In general, the vector $\boldsymbol{\theta}$ may have positive as well as negative components – so-called **short positions**. The meaning of the latter is that the agent has sold a future payment from one or more of the $N + 1$ assets although the asset is not currently in his position.

We term the vector $\boldsymbol{\theta}$ a **trading strategy**. For future reference we note:

- $\boldsymbol{\theta}$ is $\mathcal{F}_0$-measurable. Everyone would like to invest given hindsight knowledge of the realized outcome, which is not possible. Bets are made 'up front'.
- $\boldsymbol{\theta} \cdot \boldsymbol{s}_0$ is the value of the portfolio $\boldsymbol{\theta}$, i.e. the net amount of money invested.
- $\boldsymbol{\theta} \cdot \boldsymbol{S} \in \mathcal{R}^S$ is the realized payment in the different states at time T.

5.3 The 'no arbitrage' condition

Definition 1 An **arbitrage strategy** is a trading strategy $\boldsymbol{\theta} \in \mathcal{R}^{N+1}$, such that

$$\boldsymbol{\theta} \cdot [-\boldsymbol{s}_0\ ,\ \boldsymbol{S}] > 0.$$

The existence of an arbitrage opportunity can equivalently be stated as

$$\left\{\boldsymbol{y} \in \mathcal{R}^{S+1} \mid \boldsymbol{y} = \boldsymbol{\theta} \cdot [-\boldsymbol{s}_0\ ,\ \boldsymbol{S}],\ \boldsymbol{\theta} \in \mathcal{R}^{N+1}\right\} \bigcap \mathcal{R}_+^{S+1} \neq \emptyset$$

The meaning of this is that the portfolio will *never* require any net payment from the investor, although it

- either has a negative cost at time $t = 0$, reflecting the fact that the buyer of the portfolio receives money as the net result of the initial trading;
- or the portfolio position provides a positive payment in at least one possible future state s.

Both possibilities may be present. It is obvious that an opportunity to obtain something of value for nothing cannot survive for long. Everyone would rush to exploit this opportunity, creating an excess demand for some assets and an excess supply of other assets. In order to create an equality between the supply of and the demand for the assets in the capital market, prices would have to be modified in order to re-establish equilibrium in the market. We assume that this modification of prices takes place immediately, leading to the fundamental economic principle in finance that arbitrage opportunities never exist.

Definition 2 **Absence of arbitrage** means that

$$\{\boldsymbol{y} \in \mathcal{R}^{S+1} \mid \boldsymbol{y} = \boldsymbol{\theta} \cdot [-\boldsymbol{s}_0\,,\, \boldsymbol{S}]\,,\, \boldsymbol{\theta} \in \mathcal{R}^{N+1}\} \bigcap \mathcal{R}_+^{S+1} = \emptyset.$$

Absence of arbitrage has some far-reaching consequences, as we will outline in the following.

It is worth pointing out that arbitrage-based pricing theories are theories about relative prices. Theories based on 'no arbitrage' do not attempt to explain why the price of a particular stock or any particular exchange rate, say, reached their observed levels. Only the interrelationship between prices is explained. Despite the possibility that some prices may be viewed as 'incorrect', when measured against certain fundamental economic variables, arbitrage-based pricing theories will still be able to say something about the correct interrelationship between such prices.

Theorem 1 Absence of arbitrage is equivalent to the existence of a strictly positive vector $\boldsymbol{\lambda} \in \mathcal{R}_{++}^{S+1}$, such that

$$E^P\left\{[-\boldsymbol{s}_0\,,\, \boldsymbol{S}]\boldsymbol{\lambda}\right\} = 0. \tag{5.2}$$

The main tool for proving Theorem 1 is the following separation theorem for convex subsets of $\mathcal{R}^m$ (see, for example, Rockafellar, 1970, Cor. 11.4.2, p. 99).

Theorem 2 (Minkowski's separation theorem) Let C_1 and C_2 be two non-empty, disjoint convex and closed sets in $\mathcal{R}^m$. If C_2 is compact, then there exists

- a vector $\lambda \in \mathcal{R}^m$, $\lambda \neq 0$ and
- two constants $a, b \in \mathcal{R}$

such that

$$<\lambda, x>_P \equiv \sum_{i=1}^{m} P_i \lambda_i x_i \leq a < b \leq <\lambda, y>_P$$
$$\equiv \sum_{i=1}^{m} P_i \lambda_i y_i \quad \text{for all } x \in C_1,\ y \in C_2. \quad (5.3)$$

The strict inequality in (5.3) is guaranteed by the compactness of C_2.

Proof of Theorem 1 It is straightforward to prove that (5.2) excludes arbitrage opportunities.

Let $E^P\{[-s_0,\ S]\lambda\} = 0$ for $\lambda \in \mathcal{R}^{S+1}_{++}$, and assume that an arbitrage strategy θ exists. Then we have a contradiction, because

$$\theta \cdot [-s_0,\ S] > 0 \text{ and } \lambda \in \mathcal{R}^{S+1}_{++} \Rightarrow <\theta \cdot [-s_0,\ S], \lambda>_P \in \mathcal{R}_+$$

and

$$<[-s_0,\ S]\ ,\ \lambda>_P = 0 \Rightarrow <\theta \cdot [-s_0,\ S], \lambda>_P = 0.$$

In order to prove the reverse, define $\mathcal{M}$ as the linear subspace of $\mathcal{R}^{S+1}$,

$$\mathcal{M} = \left\{y \in \mathcal{R}^{S+1} \mid y = \theta \cdot [-s_0,\ S],\ \theta \in \mathcal{R}^{N+1}\right\},$$

and the unit simplex as the subset of $\mathcal{R}^{S+1}_+$:

$$\Delta^S = \{y \in \mathcal{R}^{S+1}_+ \mid \sum_{j=0}^{S} P_j y_j = 1\}.$$

Absence of arbitrage means by definition that $\mathcal{M} \bigcap \mathcal{R}^{S+1}_+ = \emptyset$. A fortiori, it is also the case that $\mathcal{M} \bigcap \Delta^S = \emptyset$. Since the sets $\mathcal{M}$ and Δ^S fulfil the conditions in Minkowski's separation theorem, there exist a vector $\lambda \in \mathcal{R}^{S+1}$ and two real constants a and b, such that

$$<\lambda, x>_P \leq a < b \leq <\lambda, y>_P \quad \forall x \in \mathcal{M},\ y \in \Delta^S. \quad (5.4)$$

The constant a must be 0 and $<\lambda, x>_P = 0 \ \ \forall x \in \mathcal{M}$. This is so, because $\mathcal{M}$ as a linear subspace contains all multiples of vectors $x \in \mathcal{M}$. Consequently, $b > 0$.

It follows that λ is a vector in $\mathcal{M}^\perp$ – with respect to the inner product $<,>_P$ – which is exactly the statement of Theorem 1. It remains to be proved that λ is a vector in $\mathcal{R}^{S+1}_{++}$.

The simplex Δ^S contains all positive unit vectors. This excludes the presence of both negative and zero components in $\boldsymbol{\lambda}$, since this would violate the separation property in (5.4) with its strict inequality. Hence, $\boldsymbol{\lambda} \in \mathcal{R}_{++}^{S+1}$.

Given that all components of the separation vector $\boldsymbol{\lambda}$ are positive, it is possible to normalize, with no loss of generality, by imposing the condition $\lambda_0 = 1$. This leads to the following corollary:

Corollary 1 Absence of arbitrage is equivalent to the existence of positive coefficients $\hat{\lambda}_\omega$ such that

$$s_0 = \sum_{\omega=1}^{S} P_\omega \hat{\lambda}_\omega s_T(\omega) = E^P\left[\boldsymbol{S}\hat{\boldsymbol{\lambda}}\right]. \tag{5.5}$$

The elements $P_\omega \hat{\lambda}_\omega$ are called **state prices**, and the vector made up of these prices is called the **state price vector**. This relates to the fact that $P_\omega \hat{\lambda}_\omega$ is the price of an asset – not necessarily an existing one, represented in matrix $\boldsymbol{S}$ – whose payoff vector is the ωth unit vector. Such assets are called **Arrow–Debreu securities.**

Example 1 Consider the following case with two assets and two states:

$$s_0 = \begin{bmatrix} 1 \\ 1 \end{bmatrix} \qquad \boldsymbol{S} = \begin{bmatrix} 1 & 1 \\ 2 & a \end{bmatrix}. \tag{5.6}$$

We try to solve (5.5) for the state prices $(P_1\hat{\lambda}_1, P_2\hat{\lambda}_2)$. For $a \neq 2$ this is possible, whereas no solution exists for $a = 2$:

$$\begin{bmatrix} 1 & 1 \\ 2 & a \end{bmatrix} \cdot \begin{bmatrix} P_1\hat{\lambda}_1 \\ P_2\hat{\lambda}_2 \end{bmatrix} = \begin{bmatrix} 1 \\ 1 \end{bmatrix} \quad \Rightarrow \quad \begin{bmatrix} P_1\hat{\lambda}_1 \\ P_2\hat{\lambda}_2 \end{bmatrix} = \begin{bmatrix} \frac{1-a}{2-a} \\ \frac{1}{2-a} \end{bmatrix}, \tag{5.7}$$

We observe the following from (5.7):

- For $a < 1$ the state prices are both positive, and it is exactly these parameter values for a that exclude arbitrage opportunities according to Corollary 1.
- For $1 \leq a$ an arbitrage opportunity exists, because the second asset dominates the first asset. The non-positivity of the state prices reveals the existence of an arbitrage opportunity. It is easy to check that any portfolio of the form $\boldsymbol{\theta} = (-1, y)$ with $\max\{\frac{1}{2}, \frac{1}{a}\} \leq y \leq 1$ results in:
  - net investment $y - 1 \leq 0$ at time 0;
  - payment $(2y - 1, ay - 1) \in \mathcal{R}_+^2$ in period 2.

- For $a = 1$, only portfolios that are positive multiples of $(-1, 1)$ create arbitrage opportunities. They can only generate a zero net investment at time 0 and a zero payment in state 2 at time T.
- For $a > 1$ it is possible to have simultaneously a negative net investment at time 0 and positive payments in both states at time T.
- For $a = 1$ there is one state price which is positive and one state price which is zero.
- For $1 < a < 2$ there is one positive and one negative state price.
- For $a = 2$ there exist no solutions to the equations for the state prices.
- For $a > 2$ both state prices are negative.

As is seen from this extremely simple example, whenever the 'no arbitrage' condition fails it is not possible in general to tell in advance where strict inequalities can occur or whether and how the existence of arbitrage opportunities affects the sign or the existence of the state prices.

We have been abusing the term 'state prices' slightly here, since by their introduction it was tacitly understood that such prices were positive. The moral of the example can also be phrased in the following way.

We are looking for a linear functional acting on the rows of S and with prescribed values given by the elements of s_0. For $a = 2$ no such functional exists, whereas for $a \neq 2$ such a linear functional exists. Moreover, it is uniquely determined. However, the functional will only be strictly positive for $a < 1$, and strict positivity is exactly what is required in order to avoid arbitrage opportunities.

5.4 Equivalent martingale measures and risk-neutralized price processes

So far we have not specified anything about the denomination of prices.

Assume that we are considering a model with agents residing in two different countries. An agent residing in country A would naturally use currency A as his unit of account, and an agent residing in country B would naturally use currency B as his unit of account. Any agent would naturally use that unit of account which is generally accepted for transactions purposes in the market in which he operates. A unit of account is also termed a **numéraire**.

If an arbitrage opportunity exists in terms of numéraire A, an arbitrage opportunity also exists whenever all prices are converted into numéraire B or into any other numéraire. The only restriction for switching between two numéraires is that they have positive prices in terms of each other at all times and in all future states. Hence, all agents reach the same conclusions concerning the consequences of absence of arbitrage, although these conclusions may be expressed in terms of different numéraires.

This observation will be further analysed in this and the following section, where we are looking for alternative ways of expressing (5.5). It turns out that the ability to switch between different numéraires enhances the ability in more complicated probability-theoretic settings both to deduce certain closed-form solutions and to construct efficient numerical routines for models that do not allow for closed-form solutions.

Assume that asset no. 0 is a riskless bond paying one unit of the numéraire currency in all states $\omega \in \Omega$ at time T. In mathematical terms this means that $s_T^0 = (1, 1, \ldots, 1)$ and that the price is $s_0^0 = \sum_{\omega=1}^{S} P_\omega \hat{\lambda}_\omega$. By definition of the **interest rate** r, this price can also be expressed as $s_0^0 = (1 + r)^{-T}$.

Given the result of Theorem 1 we can now – for any possible $\hat{\lambda}$ – define a new probability measure Q:

$$Q\{\omega\} \equiv \frac{P_\omega \hat{\lambda}_\omega}{\sum_\omega P_\omega \hat{\lambda}_\omega} = \frac{P_\omega \hat{\lambda}_\omega}{(1+r)^{-T}} > 0 \quad \text{for } \omega = 1, 2, \ldots, S. \tag{5.8}$$

Due to the fact that $\hat{\lambda} \in \mathcal{R}_{++}^{S+1}$, Q is equivalent to P. Making use of Corollary 1, prices of financial assets can be written as follows:

$$s_0^n = \sum_{\omega=1}^{S} \frac{P_\omega \hat{\lambda}_\omega}{(1+r)^{-T}} \frac{s_T^n(\omega)}{(1+r)^T} = E^Q \left[\frac{s_T^n}{(1+r)^T} \right], \tag{5.9}$$

or in terms of the Radon–Nikodym derivative:

$$\rho_\omega^T \equiv \frac{dQ}{dP}\{\omega\} = \frac{Q\{\omega\}}{P\{\omega\}} = \hat{\lambda}_\omega (1+r)^T \qquad \rho^0 \equiv E^P \left[\rho^T \mid \mathcal{F}_0 \right] = 1 \tag{5.10}$$

$$s_0^n = \sum_\omega P_\omega \left(\rho_\omega^T \frac{s_T^n(\omega)}{(1+r)^T} \right). \tag{5.11}$$

For future reference we restate this as follows:

$$\frac{s_0^n}{(1+r)^0} = E^Q \left[\frac{s_T^n}{(1+r)^T} \right] = E^P \left[\frac{\rho^T s_T^n}{(1+r)^T} \right]. \tag{5.12}$$

In words:

- The processes $s_\tau^n/(1+r)^\tau$ are Q-martingales.
- The processes $\rho^\tau s_\tau^n/(1+r)^\tau$ are P-martingales.

The measure Q is termed an **equivalent martingale measure.** The process $\rho^\tau s_\tau^n$ is termed the **risk-neutralized price process** for asset n.

These equivalent ways of expressing the prices are representation theorems. The risk-neutralized process is not one observed in actual markets, and the equivalent martingale measure does not coincide with any individual investor's perception of the world as expressed by P (unless such an investor happened to be risk-neutral). However, these concepts are useful for pricing purposes.

5.5 Change of numéraire

In section 5.4 we constructed an equivalent martingale measure by using asset no. 0 – the riskless bond – as the normalization device. More precisely, we normalized the price process to $s_0^0 = 1$ and $s_T^0 = (1+r)^T$ in order to ensure that the new measure Q was also a probability measure.

There is nothing unique about this choice of numéraire. Any asset j with $s^j \in \mathcal{R}_{++}^{S+1}$ can serve as a numéraire. Consider

$$\frac{s_0^n}{(1+r)^0} = E^Q\left[\frac{s_T^n}{(1+r)^T}\right] = E^P\left[\frac{\rho^T s_T^n}{(1+r)^T}\right] \tag{5.13}$$

$$\frac{s_0^j}{(1+r)^0} = E^Q\left[\frac{s_T^j}{(1+r)^T}\right] = E^P\left[\frac{\rho^T s_T^j}{(1+r)^T}\right]. \tag{5.14}$$

Dividing (5.13) by (5.14), we arrive at

$$\frac{s_0^n}{s_0^j} = E^Q\left[\frac{s_T^n}{s_T^j}\frac{s_T^j}{E^Q\left(s_T^j\right)}\right] = E^P\left[\frac{\rho^T s_T^n}{\rho^T s_T^j}\frac{\rho^T s_T^j}{E^P\left(\rho^T s_T^j\right)}\right]. \tag{5.15}$$

Applying $\frac{s_T^j}{E^Q\left(s_T^j\right)}$ as a Radon–Nikodym derivative, we can restate (5.15) as

$$\frac{s_0^n}{s_0^j} = E^{Q_j}\left[\frac{s_T^n}{s_T^j}\right], \tag{5.16}$$

where Q^j is also a probability measure, equivalent to P and Q. Hence, for every numéraire with positive prices in all states, there exists a probability measure with the property that all prices, expressed in this numéraire, become martingales.

5.6 Attainability

A vector $\boldsymbol{\delta}(\boldsymbol{\theta}) \in \mathcal{R}^S$ is said to be **attainable**, if there exists a trading strategy $\boldsymbol{\theta}$ so that $\boldsymbol{\theta} \cdot \boldsymbol{S} = \boldsymbol{\delta}(\boldsymbol{\theta})$. The set of all attainable vectors is a linear subspace of $\mathcal{R}^S$ – the subspace spanned by the rows of $\boldsymbol{S}$. It is also the projection of $\mathcal{M} \subseteq \mathcal{R}^{S+1}$ onto $\mathcal{R}^S$, where $\mathcal{M}$ was defined in the proof of Theorem 1. For this reason we denote this space as $\mathcal{M}_0$.

Definition 3 A **price system** is a linear functional $\pi : \mathcal{M}_0 \to \mathcal{R}$ which is strictly increasing. A price system is said to be consistent with $(\boldsymbol{s}_0, \boldsymbol{S})$, if $\pi(\boldsymbol{\theta} \cdot \boldsymbol{S}) = \boldsymbol{\theta} \cdot \boldsymbol{s}_0 \ \forall \boldsymbol{\theta} \in \mathcal{R}^{N+1}$.

The set of price systems consistent with $(\boldsymbol{s}_0, \boldsymbol{S})$ is called $\boldsymbol{\Pi}$.

The existence of a consistent price system implies the absence of arbitrage. This is so because $\boldsymbol{\theta} \cdot \boldsymbol{S} \in \mathcal{R}^S_{\mp}$ implies $\boldsymbol{\theta} \cdot \boldsymbol{s}_0 > 0$ by the requirement of π being strictly increasing. Hence, the defining expression can be continued in the following way:

$$\begin{aligned}\pi(\boldsymbol{\theta} \cdot \boldsymbol{S}) &= \boldsymbol{\theta} \cdot \boldsymbol{s}_0 = \boldsymbol{\theta} \cdot E^P\left[\boldsymbol{S}\hat{\boldsymbol{\lambda}}\right] = \boldsymbol{\theta} \cdot E^Q\left[\boldsymbol{S}(1+r)^{-T}\right] \\ &\quad \forall \boldsymbol{\theta} \in \mathcal{R}^{N+1}\ \forall \hat{\boldsymbol{\lambda}}. \end{aligned} \tag{5.17}$$

Any positive linear functional π defined on the subspace $\mathcal{M}_0$ can be extended to a positive linear functional $\psi : \mathcal{R}^S \to \mathcal{R}$ in such a manner that $\psi(\boldsymbol{\delta}(\boldsymbol{\theta})) = \pi(\boldsymbol{\delta}(\boldsymbol{\theta}))$ for all $\boldsymbol{\delta} \in \mathcal{M}_0$. The set of all such consistently extended functionals is called $\boldsymbol{\Psi}$.

We assume – as in section 5.4 – that asset no. 0 is a riskless asset. Consider now a class of probability measures, denoted by $\mathcal{Q}$, whose elements satisfy the martingale property

$$Q \in \mathcal{Q} \quad \Leftrightarrow \quad \boldsymbol{\theta} \cdot \boldsymbol{s}_0 = E^Q\left[\boldsymbol{\delta}(\boldsymbol{\theta}(1+r)^{-T}\right]. \tag{5.18}$$

From the derivation in sections 5.3 and 5.4 we know that excluding any arbitrage strategy is sufficient to guarantee that $\mathcal{Q}$ contains at least one element. The elements of $\mathcal{Q}$ can, however, also be characterized in terms of the linear functionals in $\boldsymbol{\Psi}$:

Theorem 3 There is a one-to-one correspondence between $\boldsymbol{\Psi}$ and $\mathcal{Q}$ given by

a. $\psi[\boldsymbol{X}] = E^Q\left[\boldsymbol{X}(1+r)^{-T}\right] \quad \forall \boldsymbol{X} \in \mathcal{R}^S$.

b. $Q[A] = (1+r)^T \psi[1_A] \quad \forall A \in \mathcal{F}$.

Proof If $Q \in \mathcal{Q}$, the functional ψ defined by (a) will be a linear functional. ψ will also be positive, because

$$\boldsymbol{X}^* \in \mathcal{R}^S_{\mp} \Rightarrow \boldsymbol{X}^*(1+r)^{-T} \in \mathcal{R}^S_{\mp} \Rightarrow E^Q[\boldsymbol{X}^*(1+r)^{-T}] > 0,$$

due to the equivalence of the measures P and Q.

Hence, $\psi[\boldsymbol{X}^*] > 0$, and ψ is a price system. The consistency follows immediately from the fact that $\boldsymbol{\delta}(\boldsymbol{\theta})(1+r)^{-T} > 0$ is a martingale under any measure $Q \in \mathcal{Q}$.

Let ψ be a consistent price system. By assumption there are no null sets in 2^{Ω}. And $1_A \in \mathcal{R}^S_+$ implies $\psi\{1_A\} > 0$. Furthermore, $\psi\{1_\Omega\} = (1+r)^{-T}$, so $Q(\Omega) = 1$. Hence, the measures P and Q are equivalent probability measures.

It remains to prove that $\delta(\theta)(1+r)^{-T}$ is a Q-martingale for $\delta(\theta) \in \mathcal{M}_0$. However, by linearity of ψ we have

$$E^Q(\delta(\theta)(1+r)^{-T}) = \sum_{\omega=1}^{S} \delta(\theta)_\omega (1+r)^{-T}(1+r)^T \psi(1_{\{\omega\}}) =$$

$$\sum_{\omega=1}^{S} \delta(\theta)_\omega \psi(1_{\{\omega\}}) = \psi\left(\sum_{\omega=1}^{S} \delta(\theta)_\omega 1_{\{\omega\}}\right) = \theta \cdot s_0 \tag{5.19}$$

where the latter equality follows from the consistency requirement.

Finally, if Q_1 and Q_2, respectively, are two martingale measures from the set $\mathcal{Q}$, and ψ_1 and ψ_2 are the corresponding extended functionals in Ψ, then $\psi_1 = \psi_2$ implies that

$$Q_1(\{\omega\}) = (1+r)^T \psi_1(1_{\{\omega\}}) = (1+r)^T \psi_2(1_{\{\omega\}}) = Q_2(\{\omega\}) \quad \forall\, \omega \in \Omega. \tag{5.20}$$

Hence, the correspondence is one-to-one.

Definition 4 The market is said to be **complete**, if $\mathcal{M}_0 = \mathcal{R}^S$.

It is obvious that this definition implies that $\Psi = \{\psi\}$ and $\mathcal{Q} = \{Q\}$.

The prime reason why this concept is of interest is that in a complete market, any asset with a $\mathcal{F}_T$-measurable payment at time T has a unique price consistent with 'no arbitrage'. This is stated in Theorem 4 below. The assets in question include not only the existing ones listed in s_0, but also any new asset that might be introduced, such as options.

This statement needs some qualifying comments. The basic state space Ω and the associated filtration do not necessarily specify anything of economic relevance. Rather, they are usually constructed 'backwards' in order to provide the simplest possible probabilistic set-up capable of describing the price processes in an adequate manner.

Example 2 Consider the following simple example with $S = 3$, $N = 2$:

ω	$s^0_T(\omega)$	$s^1_T(\omega)$	$s^2_T(\omega)$	$P(\omega)$	$Q(\omega)$
1	1	2	1	$\frac{1}{4}$	$\frac{1}{3}$
2	1	1	2	$\frac{1}{2}$	$\frac{1}{3}$
3	1	2	2	$\frac{1}{4}$	$\frac{1}{3}$

In this example the market is complete. However, in order to describe the marginal distributions of assets 1 and 2, a state space of cardinality 2 is sufficient:

ω	$s_T^0(\omega)$	$s_T^1(\omega)$	$P(\omega)$	$Q(\omega)$
1	1	2	$\frac{1}{2}$	$\frac{2}{3}$
2	1	1	$\frac{1}{2}$	$\frac{1}{3}$

ω	$s_T^0(\omega)$	$s_T^2(\omega)$	$P(\omega)$	$Q(\omega)$
1	1	2	$\frac{3}{4}$	$\frac{2}{3}$
2	1	1	$\frac{1}{4}$	$\frac{1}{3}$

A **contingent claim** with one asset as so-called **underlying asset** is a random payment measurable with respect to the σ-algebra generated by the price process of this underlying asset. Hence, the random payment from a contingent claim with s^1 as underlying asset is given as $h(s^1)$ for some Borel function h (and analogously for contingent claims with s^2 as underlying asset). Since the assets $1_{\{s_T^1=1\}}$ and $1_{\{s_T^1=2\}}$ both have prices that are uniquely determined by the 'no arbitrage' condition without any consideration of s_T^2, any contingent claim with asset no. 1 as underlying asset is uniquely priced within a state space of cardinality 2.

A contingent claim with more than one asset as underlying assets – so-called 'multiple contingencies' – is a random payment measurable w.r.t. the σ-algebra generated by the price processes of the underlying assets. Hence, the random payment from a contingent claim with, say, (s^1, s^2) as underlying assets is given as $h(s^1, s^2)$ for some Borel function h. As an example, consider the asset $1_{\{s_T^1=1, s_T^2=2\}}$ with two underlying assets. This can only be priced by a representation with cardinality 3.

Theorem 4 The market is complete

a. if and only if rank(S)=S

b. if and only if the set Ψ is a singleton

c. if and only if the set $\mathcal{Q}$ is a singleton

d. if and only if the state price vector with elements $P_\omega \hat{\lambda}_\omega$ is unique.

Proof The equivalence of completeness and (a) is trivial, since the statement in (a) is merely a restatement of the definition.

By Theorem 3, (b) and (c) are equivalent. By construction of equivalent martingale measures there is a one-to-one correspondence between normalized separation vectors $\hat{\lambda}$ and the probability measures in $\mathcal{Q}$. Hence, (c) and (d) are equivalent statements. The theorem is proved if completeness implies (b) and (d) implies completeness.

Observe that any unit vector $e_j \in \mathcal{R}^S$ is attainable as $\theta(j) \cdot S$ for an appropriately chosen trading strategy $\theta(j)$. Hence, $\pi(e_j)$ is given as $\theta(j) \cdot s_0$. Since any attainable vector is a unique linear combination of the vectors e_j, and all vectors are attainable, the price of any trading strategy is uniquely determined. This proves that completeness implies (b).

Since $(1, \hat{\lambda}) \in \mathcal{M}^{\perp}$, the uniqueness of the state price vector assumed in (d) is equivalent to the statement that $\mathcal{M}^{\perp}$ has dimension 1. Hence, $\mathcal{M}$ has dimension S. Since s_0 is a unique combination of the columns $s_T(\omega)$, $\mathcal{M}_0$ also has dimension S. This proves that (d) implies completeness.

5.7 Multiperiod model in discrete time

The basic difference between this and the previous sections is the specification of $\mathcal{T}$. Instead of consisting of two time indices, now and one future trading date, it is generalized to $\mathcal{T} = \{0, 1, 2, \ldots, T\}$. A state in Ω is now a realization as a function of time of the processes in question.

The time indices refer to calendar times, where trading of financial assets takes place. The filtration is taken to be the filtration generated by the traded assets

$$\mathcal{F}_t = \sigma\{s_j^0, s_j^1, \ldots, s_j^N\} \quad 0 \le j \le t,$$

so that any price process is automatically an adapted process. (In general the filtration must be required to be such that the price processes are adapted.)

A trading strategy is now $(N + 1)$-dimensional process $\{\theta_t\}_{\{t=1,2,\ldots,T\}}$. θ_t^i denotes the number of units of asset i bought by an investor at time $t-1$ – given $\mathcal{F}_{t-1}$ – and held until time t. In the vocabulary of stochastic processes, θ_t is a predictable process.

The portfolio value at time $t-1$ is $\theta_t \cdot s_{t-1} = \sum_{i=0}^{N} \theta_t^i \cdot s_{t-1}^i$. At time t, the portfolio can be liquidated for the value $\theta_t \cdot s_t$, and the proceeds can be reinvested in a new portfolio θ_{t+1} based on $\mathcal{F}_t$.

The set of all trading strategies is denoted by Θ. In this paper we assume that the traded assets do not pay out dividends, coupon payments and similar intermediate payments to the owner, but that the return from investing in these assets consists only of capital gains. This is no real restriction, but it alleviates the notational burden considerably.

The **cash flow process** δ derived from the trading strategy θ is defined in the following manner:

$$\delta_0(\theta) = -\theta_1 \cdot s_0 \qquad (5.21)$$

$$\delta_1(\theta) = (\theta_1 - \theta_2) \cdot s_1 \tag{5.22}$$

$$\vdots$$

$$\delta_{T-1}(\theta) = (\theta_{T-1} - \theta_T) \cdot s_{T-1} \tag{5.23}$$

$$\delta_T(\theta) = \theta_T \cdot s_T. \tag{5.24}$$

This process is by construction adapted for any trading strategy $\theta \in \Theta$.

The cash flow process is a vector in $\mathcal{R} \times \mathcal{R}^S \times \mathcal{R}^T$, which is equipped with the inner product

$$E^P \left[\sum_{t=0}^{T} x_t y_t \right].$$

Definition 5

- An arbitrage strategy denotes a trading strategy $\theta \in \Theta$, whose cash flow process $\{\delta_t(\theta)\}_{t \in \mathcal{T}}$ fulfils
$$P\{\delta_t(\theta) \geq 0\} = 1 \quad \forall t \in \mathcal{T}$$
$$P\{\delta_t(\theta) > 0\} > 0 \quad \text{for at least one } t \in \mathcal{T}.$$
- Let $\mathcal{L}$ denote the set of all processes that are adapted to the filtration $\{\mathcal{F}_\tau\}_{\tau \in \mathcal{T}}$. $\mathcal{L}$ is a linear subspace of $\mathcal{R} \times \mathcal{R}^S \times \mathcal{R}^T$.
- $\mathcal{M}$ denotes the set of all cash flow processes generated by the use of trading strategies $\theta \in \Theta$. $\mathcal{M}$ is a linear subspace of $\mathcal{L}$.
- $\mathcal{L}_0$ and $\mathcal{M}_0$ denote the projection of $\mathcal{L}$ and $\mathcal{M}$, respectively, onto $\mathcal{R}^S \times \mathcal{R}^T$.

Theorem 5 Absence of arbitrage is equivalent to the existence of a stochastic process $\lambda \in \mathcal{L} \cap \mathcal{R}_{++} \times \mathcal{R}^S_{++} \times \mathcal{R}^T_{++}$ and a real constant b such that

$$E^P \left[\sum_{t=0}^{T} \lambda_t \delta_t(\theta) \right] = 0 < b \leq E^P \left[\sum_{t=0}^{T} \lambda_t y_t \right]$$
$$\forall\, \delta(\theta) \in \mathcal{M},\ y \in \Delta^{|\mathcal{L}|}. \tag{5.25}$$

Absence of arbitrage is equivalent to the statement that the unit simplex $\Delta^{|\mathcal{L}|}$ in $\mathcal{L}$ does not intersect the subspace $\mathcal{M}$. Hence, the proof of theorem 5 is entirely analogous to that of Theorem 1. Since λ is a strictly positive and adapted process, we can once again normalize by imposing the condition $\lambda_0 = 1$ and designate this particular choice of separation vector as $\hat{\lambda}$. Writing out the statement in (5.25) we have the following corollary.

Corollary 2 Absence of arbitrage is equivalent to the existence of a positive and adapted stochastic process $\hat{\lambda}$ such that

$$\theta_1 \cdot s_0 = E^P\left[\sum_{t=1}^{T} \hat{\lambda}_t \delta_t(\theta)\right], \tag{5.26}$$

for any trading strategy $\theta \in \Theta$.

Numerous conclusions can be drawn from Corollary 2.

Consider first the class of trading strategies that are designed in order to buy a particular asset, no. i, at a certain point in time, t_1, and keep it until a future point in time, t_2. This includes the case $t_1 = 0$ and/or $t_2 = T$:

$$\begin{aligned} \theta_t^k &= 0 && \forall t\, \forall k \neq i \\ \theta_t^i &= 0 && \text{for } t \leq t_1 \\ \theta_t^i &= 1 && \text{for } t_1 < t \leq t_2 \\ \theta_t^i &= 0 && \text{for } t_2 < t \leq T. \end{aligned}$$

For any such strategy we can write equation (5.26) which specializes to:

$$E^P\left[\hat{\lambda}_{t_1} s_{t_1}^i\right] = E^P\left[\hat{\lambda}_{t_2} s_{t_2}^i\right]. \tag{5.27}$$

For $t_1 = 0$ (5.27) becomes

$$s_0^i = E^P\left[\hat{\lambda}_{t_2} s_{t_2}^i\right]. \tag{5.28}$$

Let $t_2 = T$ and consider an asset j whose price process is assumed to be strictly positive. From (5.28) we have that for any such asset j

$$\rho_j^T = \frac{\hat{\lambda}_T s_T^j}{s_0^j}, \tag{5.29}$$

defines a positive random variable on $(\Omega, \mathcal{F})$ with P-mean equal to 1. Hence, it is a Radon–Nikodym derivative. And

$$\frac{s_0^i}{s_0^j} = \sum_{\omega=1}^{S}\left[P_\omega \frac{\hat{\lambda}_T(\omega) s_T^j(\omega)}{s_0^j}\right] \frac{s_T^i(\omega)}{s_T^j(\omega)} \equiv E^{Q_j}\left[\frac{s_T^i}{s_T^j}\right]. \tag{5.30}$$

where $\frac{dQ_j}{dP} = \rho_j^T$.

The measure Q_j is equivalent to P. It remains to prove, however, that Q_j is an equivalent martingale measure, i.e.

$$\frac{s_t^i}{s_t^j} = E^{Q_j}\left[\frac{s_T^i}{s_T^j} \mid \mathcal{F}_t\right]. \tag{5.31}$$

Consider the following generalized class of trading strategies, where A is a particular, but arbitrary, set in $\mathcal{F}_{t_1}$:

$$\begin{aligned} \theta_t^k &= 0 && \forall t\, \forall k \neq j \\ \theta_t^j &= 0 && \text{for } t \leq t_1 \\ \theta_t^j &= 1 && \text{for } t_1 < t \leq T \ \text{ for } \omega \in A \\ \theta_t^j &= 0 && \text{for } t_1 < t \leq T \ \text{ for } \omega \notin A. \end{aligned}$$

The statement in (5.27) can now be rephrased as

$$E^P\left[\frac{\hat{\lambda}_{t_1} s_{t_1}^j}{s_0^j} 1_A\right] = E^P\left[\frac{\hat{\lambda}_T s_T^j}{s_0^j} 1_A\right] = E^{Q_j}[1_A] = Q_j\{A\} \qquad \forall A \in \mathcal{F}_{t_1}. \tag{5.32}$$

Hence,

$$\frac{\hat{\lambda}_{t_1} s_{t_1}^j}{s_0^j} = E^P\left[\frac{\hat{\lambda}_T s_T^j}{s_0^j} \mid \mathcal{F}_{t_1}\right] = E^P\left[\rho_j^T \mid \mathcal{F}_{t_1}\right]. \tag{5.33}$$

Analogously, the same relation is valid for asset i:

$$\frac{\hat{\lambda}_{t_1} s_{t_1}^i}{s_0^j} = E^P\left[\frac{\hat{\lambda}_T s_T^i}{s_0^j} \mid \mathcal{F}_{t_1}\right]. \tag{5.34}$$

Combining (5.33) and (5.34) leads to:

$$\frac{\hat{\lambda}_{t_1} s_{t_1}^j}{s_0^j} \frac{s_{t_1}^i}{s_{t_1}^j} = E^P\left[\rho_j^T \frac{s_T^i}{s_T^j} \mid \mathcal{F}_{t_1}\right] \Rightarrow$$

$$\frac{s_{t_1}^i}{s_{t_1}^j} = \frac{E^P\left[\rho_j^T \frac{s_T^i}{s_T^j} \mid \mathcal{F}_{t_1}\right]}{E^P\left[\rho_j^T \mid \mathcal{F}_{t_1}\right]} = E^{Q_j}\left[\frac{s_T^i}{s_T^j} \mid \mathcal{F}_{t_1}\right], \tag{5.35}$$

and the result follows.

The strategies described so far were specific to single assets as opposed to portfolios of assets. However, they immediately generalize to the portfolio values of any trading strategy. A particular useful class of trading strategies are those satisfying the **self-financing** condition: The cash flow process is identically zero for $t > 0$ until a liquidation date occurs. This property is invariant with respect to the choice of numéraire. Zero is zero independent of the numéraire.

By a straightforward generalization of (5.35), the portfolio values of any self-financing trading policy have the martingale property. We state this explicitly as a theorem:

Theorem 6 Consider a trading strategy θ satisfying the self-financing condition up to a certain liquidation date t_2:

$$\begin{aligned} \delta_t(\theta) &= 0 \ \forall 0 < t < t_2 && (5.36) \\ \delta_{t_2}(\theta) &= \theta_{t_2} \cdot s_{t_2}. && (5.37) \end{aligned}$$

When expressed in terms of a numéraire with a strictly positive price process, s^j, the portfolio values

$$\sum_{i=0}^{N} \theta_t^i \cdot \frac{s_t^i}{s_t^j}, \quad 0 \le t \le t_2, \quad \text{with } \theta_0^i \equiv \theta_1^i, \tag{5.38}$$

form a Q_j-martingale.

Proof Consider the identity:

$$\frac{\theta_{t_2} \cdot s_{t_2}}{s_{t_2}^j} = \frac{\theta_0 \cdot s_0}{s_0^j} + \sum_{v=1}^{t_2} \theta_v \cdot \left(\frac{s_v}{s_v^j} - \frac{s_{v-1}}{s_{v-1}^j} \right) + \sum_{v=1}^{t_2} (\theta_v - \theta_{v-1}) \cdot \frac{s_{v-1}}{s_{v-1}^j}. \tag{5.39}$$

The last summation is zero by assumption. A sequential application of conditional expectations shows that the middle term is also zero. The first step is

$$\begin{aligned} E^{Q_j}\left[\theta_{t_2} \cdot \frac{s_{t_2}}{s_{t_2}^j} \middle| \mathcal{F}_{t_2-1} \right] &= \frac{\theta_0 \cdot s_0}{s_0^j} \\ &+ \theta_{t_2} \cdot \left(E^{Q_j}\left[\frac{s_{t_2}}{s_{t_2}^j} \middle| \mathcal{F}_{t_2-1} \right] - \frac{s_{t_2-1}}{s_{t_2-1}^j} \right) \\ &+ \sum_{v=1}^{t_2-1} \theta_v \cdot \left(\frac{s_v}{s_v^j} - \frac{s_{j-1}}{s_{v-1}^j} \right) \\ &= \frac{\theta_0 \cdot s_0}{s_0^j} + \sum_{v=1}^{t_2-1} \theta_v \cdot \left(\frac{s_v}{s_v^j} - \frac{s_{j-1}}{s_{v-1}^j} \right) \\ &= \frac{\theta_{t_2-1} \cdot s_{t_2-1}}{s_{t_2-1}^j}. \end{aligned}$$

This sequential conditioning procedure stops when conditioning with $\mathcal{F}_0$, where we are left with the first term.

In the rest of this section we will use asset 0 as the numéraire and simply denote the associated equivalent martingale measure by Q. We let $s_0^0 \equiv 1$. The symbol ˜ will denote prices and values measured according to this numéraire asset:

$$\tilde{s}_t^i \equiv \frac{s_t^i}{s_t^0} \qquad\qquad \tilde{\delta}_t(\theta) \equiv \frac{\delta_t(\theta)}{s_t^0}.$$

Definition 6 A **consistent price system** is a strictly positive linear functional π, defined on $\mathcal{M}_0$, satisfying the requirement

$$\pi(\delta(\theta)) = \theta_1 \cdot s_0.$$

Given the 'no arbitrage' condition we have already established the relation

$$\pi(\delta(\theta)) = E^Q\left[\sum_{t=1}^{T} \tilde{\delta}_t(\theta)\right] \qquad \forall\theta \in \Theta. \tag{5.40}$$

We will now construct a positive extension of π to the vector space $\mathcal{L}_0$. We will show that there is a bijection between the set of equivalent martingale measures $\mathcal{Q}$ and the possible positive extensions.

Theorem 7 There exists a bijection between positive linear functionals $\psi : \mathcal{L}_0 \to R$ that are consistent extensions of π and $\mathcal{Q}$ given by

a. $\psi(\delta) := E^Q\left[\sum_{t=1}^{T} \tilde{\delta}_t\right] \quad \forall\delta \in \mathcal{L}_0$

b. Q(A):=$\psi(0, 0, \ldots, 0, s_T^0 1_A) \quad \forall A \in \mathcal{F}_T$.

Proof ψ is positive, because the expectation of a random variable with values in $\mathcal{R}_+^S$ is positive, whenever Q is equivalent to P. Since π and ψ coincide on the subspace $\mathcal{M}_0$, and the measure Q is given, the definition of ψ in (a) is an extension of π to $\mathcal{L}_0$.

Since ψ is a positive linear functional, Q is a positive measure. For $A = \Omega$ we have $Q(A)$=1 due to the assumed normalization $s_0^0 = 1$ of the numéraire asset. It remains to prove that the measure Q defined by (b) is in fact a martingale measure. To prove this it is sufficient to prove that

$$E^Q\left[X\tilde{s}_t^i\right] = E^Q\left[X\tilde{s}_{t-1}^i\right] \qquad \forall X\ \mathcal{F}_{t-1}\text{-measurable.} \tag{5.41}$$

This is an immediate consequence of the following self-financing trading strategy.

time τ	θ^0_τ	θ^i_τ
$\tau = 1$	0	0
$\vdots$	$\vdots$	$\vdots$
$\tau = t-1$	0	0
$\tau = t$	$-X\tilde{s}^i_{t-1}$	X
$\tau = t+1$	$X(\tilde{s}^i_t - \tilde{s}^i_{t-1})$	0
$\vdots$	$\vdots$	$\vdots$
$\tau = T$	$X(\tilde{s}^i_t - \tilde{s}^i_{t-1})$	0

This trading strategy generates the cash flow process

$$\begin{aligned} \tilde{\delta}_t(\theta) &= 0 \quad \text{for } 0 \leq t < T \\ \tilde{\delta}_T(\theta) &= X(\tilde{s}^i_t - \tilde{s}^i_{t-1}). \end{aligned}$$

Since ψ is consistent, $\psi(0, 0, \ldots, \delta_T(\theta)) = 0$. By the definition of measure Q as given in (b) we have

$$E^Q\left[X(\tilde{s}^i_t - \tilde{s}^i_{t-1})\right] = 0$$

which proves (5.41).

5.8 Complete market and splitting index

Definition 7 The market is said to be complete if any cash flow process can be realized by a trading strategy, i.e. if $\mathcal{M}_0$=$\mathcal{L}_0$.

Theorem 8 Given the absence of arbitrage, the market is complete if and only if there exists a unique equivalent martingale measure associated with any numéraire with a strictly positive price process.

Proof Given the absence of arbitrage we know that $\mathcal{Q} \neq \emptyset$.

Assume that the market is complete. Let Q_1 and Q_2 be two equivalent martingale measures associated with the numéraire s^0.

Let $A \in \mathcal{F}_T$ and consider the trading strategy θ giving rise to the cash flow process

$$\delta_T(\theta) = s^0_T \cdot 1_A \qquad \delta_t(\theta) = 0 \quad \text{for } 0 < t < T.$$

This trading strategy gives the pricing relations

$$\theta_1 \cdot s_0 = E^{Q_1}(1_A) = E^{Q_2}(1_A).$$

Since A was an arbitrary set in $\mathcal{F}_T$, $Q_1 = Q_2$.

If the market is not complete, $\mathcal{M}_0$ is a proper subspace of $\mathcal{L}_0$. Hence, there exists a vector $\boldsymbol{X} \in \mathcal{L}_0$ such that

$$\boldsymbol{X} \neq \boldsymbol{0} \qquad \text{and} \qquad E^Q\left[\sum_{t=1}^{T} \boldsymbol{X}_t \tilde{\delta}_t(\boldsymbol{\theta})\right] = 0 \quad \forall \tilde{\delta} \in \mathcal{M}_0. \tag{5.42}$$

We show first that $\boldsymbol{X}_t$ is a martingale. Let $A \in \mathcal{F}_{t_1}$ and consider the trading strategies

$$\begin{array}{lcll} \theta_t^k & = & 0 & \forall t\, \forall k \neq 0 \\ \theta_t^0 & = & 0 & \text{for } t \leq t_1 \\ \theta_t^0 & = & 1 & \text{for } t_1 < t \leq T \text{ and } \omega \in A \\ \theta_t^0 & = & 0 & \text{for } t_1 < t \leq T \text{ and } \omega \notin A. \end{array}$$

The cash flow process derived from this is

$$\begin{array}{lcll} \tilde{\delta}_t(\boldsymbol{\theta}) & = & 0 & \forall\, t \neq t_1, T \\ \tilde{\delta}_{t_1}(\boldsymbol{\theta}) & = & -1_A & \\ \tilde{\delta}_T(\boldsymbol{\theta}) & = & 1_A. & \end{array}$$

Writing out (5.42) leads to

$$E^Q\left[1_A X_{t_1}\right] = E^Q\left[1_A X_T\right] \qquad \forall\, A \in \mathcal{F}_{t_1},$$

showing that $\boldsymbol{X}_t$ is a martingale. Additionally, this rules out the possibility that $\boldsymbol{X}_T = \boldsymbol{0}$, because then $\boldsymbol{X} = \boldsymbol{0}$ by the martingale property. Hence, we can define a new measure Q^* by

$$Q^*(\{\omega\}) \equiv \rho_X^T(\omega) \cdot Q(\{\omega\}), \qquad \rho_X^T(\omega) = 1 + \frac{X_T(\omega)}{2 \cdot \|x\|_\infty},$$

$$\|x\|_\infty = \max_{\omega \in \Omega} |X_T(\omega)|\,. \tag{5.43}$$

ρ_X^T is a Radon–Nikodym derivative because

- the normalization ensures it is a positive random variable
- the cash flow process

$$\begin{array}{lcll} \tilde{\delta}_t & = & 0 & \text{for } 1 \leq t < T \\ \tilde{\delta}_T & = & 1, & \end{array}$$

  is in $\mathcal{M}_0$, implying that $E^Q\left[\rho_X^T\right] = 1$.

To prove that Q^* is a martingale measure, it is sufficient to consider self-financing trading strategies $\boldsymbol{\theta}$ and their associated cash flow processes $\tilde{\delta}_t(\boldsymbol{\theta})$:

$$E^{Q^*}\left[\tilde{\delta}_T(\boldsymbol{\theta}) \mid \mathcal{F}_t\right] = E^Q\left[\tilde{\delta}_T(\boldsymbol{\theta}) \mid \mathcal{F}_t\right] + E^Q\left[\frac{X_T}{2 \cdot \|x\|_\infty} \tilde{\delta}_T(\boldsymbol{\theta}) \mid \mathcal{F}_t\right] =$$

$$E^Q\left[\tilde{\delta}_T(\boldsymbol{\theta}) \mid \mathcal{F}_t\right] = \boldsymbol{\theta}_t \cdot \tilde{\boldsymbol{s}}_t. \tag{5.44}$$

The second equality can be shown by exactly the same arguments as those ones applied in deducing the consequences of Corollary 2, in particular Theorem 7 and (5.41).

Completeness can also be described by means of the partition generated by the filtration $\{\mathcal{F}_t\}_{t\in\mathcal{T}}$. The partition is denoted by $\{f_t\}_{t\in\mathcal{T}}$ and f_t is the collection of atoms of the σ−algebra $\mathcal{F}_t$. An atom $A \in \mathcal{F}_t$ is a set with the property that

$$B \in \mathcal{F}_t \wedge B \subseteq A \quad \Rightarrow \quad B = \emptyset \vee B = A.$$

Let $A(t,\omega)$ denote the atom in $\mathcal{F}_t$ that contains the element $\omega \in \Omega$. The atoms in f_t split into atoms of f_{t+1} depending upon the way new information is created. The **splitting function**, denoted by $\nu(t,\omega)$, counts the number of atoms in f_{t+1} that are subsets of $A(t,\omega) \in f_t$.

Consider the matrices $\boldsymbol{S}(t,\omega)$ whose rows are made up of the prices of the $N+1$ assets in each of the possible atoms in f_{t+1} succeeding $A(t,\omega)$. Thus $\boldsymbol{S}(t,\omega)$ has $\nu(t,\omega)$ columns. Since the price processes are adapted, the market is complete if and only if rank($\boldsymbol{S}(t,\omega)$) $= \nu(t,\omega)$ for $t = 1,2,\ldots,T$ and $\omega \in \Omega$. This is so because the cash flow processes of the form

$$\delta_t = 0 \ \text{ for } t \neq \hat{t} \qquad \delta_{\hat{t}} = 1_A \ \text{ for } A \in f_{\hat{t}} \qquad t,\hat{t} \in \{1,2,\ldots,T\} \tag{5.45}$$

span $\mathcal{L}_0$. And these cash flow processes can be attained if and only if rank($\boldsymbol{S}(t,\omega)$) $= \nu(t,\omega)$ for $t = 1,2,\ldots,T$ and $\omega \in \Omega$.

The following sequence of examples shows situations with a complete market and no arbitrage, a complete market and an arbitrage opportunity, and finally an incomplete market with no arbitrage opportunity.

Example 3 In this example $S = 4$ and $N = 1$. The price processes evolve according to the following tree:

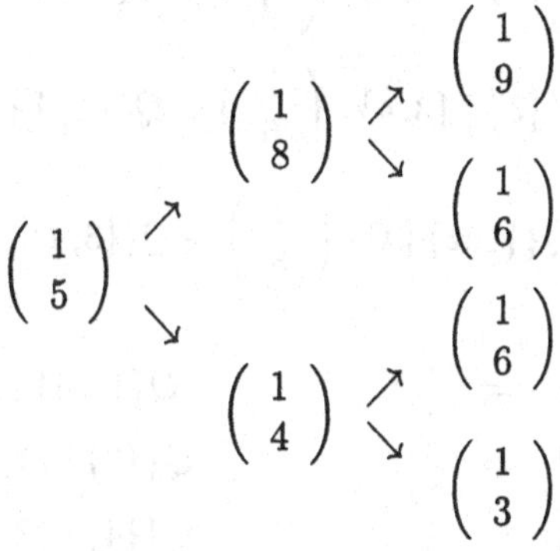

Denoting $\omega = 1$ as the 'up-up' path, $\omega = 2$ as the 'up-down' path, $\omega = 3$ as the 'down-up' path and $\omega = 4$ as the 'down-down' path, we can determine a martingale measure Q by inspection:

ω	1	2	3	4
$Q\{\omega\}$	$\frac{1}{6}$	$\frac{1}{12}$	$\frac{1}{4}$	$\frac{1}{2}$

The filtration and the partition given by the the price processes are as follows:

$$\mathcal{F}_0 = \{\emptyset, \Omega\} \qquad \mathcal{F}_1 = \{\emptyset, \Omega, \{1,2\}, \{3,4\}\} \qquad \mathcal{F}_2 = 2^\Omega$$

$$f_0 = \Omega \qquad f_1 = \{\{1,2\}, \{3,4\}\} \qquad f_2 = \{\{1\}, \{2\}, \{3\}, \{4\}\}.$$

In this example, we have splitting index 2 everywhere and no arbitrage opportunities, because a martingale measure exists. This measure is unique, revealing the completeness of the market.

On the other hand, if we just link the spot prices $(1, 5)$ at time $t = 0$ with the four possible realizations at time $t = 2$, a two-dimensional multiplicity of probabilities will fulfil the pricing equation

$$\begin{pmatrix} 1 \\ 5 \end{pmatrix} = x \begin{pmatrix} 1 \\ 9 \end{pmatrix} + (y+z) \begin{pmatrix} 1 \\ 6 \end{pmatrix} + w \begin{pmatrix} 1 \\ 3 \end{pmatrix}. \tag{5.46}$$

The solutions to (5.46) are $0 < x < \frac{1}{3}$, $y + z = \frac{2}{3} - 2x$ and $w = \frac{1}{3} + x$ for positive numbers y and z. Although the price processes behind are unchanged, the filtration is changed by removing the trading date $t = 1$. If we require the pricing at time $t = 1$ to be in accordance with the martingale measure, this adds the restrictions necessary in order to determine the measure Q uniquely by determining all the 'one-period-ahead' conditional probabilities:

$$\begin{aligned}
\begin{pmatrix} 1 \\ 8 \end{pmatrix} &= Q\{\{1\} \mid \{1,2\}\} \begin{pmatrix} 1 \\ 9 \end{pmatrix} + Q\{\{2\} \mid \{1,2\}\} \begin{pmatrix} 1 \\ 6 \end{pmatrix} \\
\begin{pmatrix} 1 \\ 4 \end{pmatrix} &= Q\{\{3\} \mid \{3,4\}\} \begin{pmatrix} 1 \\ 6 \end{pmatrix} + Q\{\{4\} \mid \{3,4\}\} \begin{pmatrix} 1 \\ 3 \end{pmatrix} \\
\begin{pmatrix} 1 \\ 5 \end{pmatrix} &= Q\{\{1,2\} \mid \Omega\} \begin{pmatrix} 1 \\ 8 \end{pmatrix} + Q\{\{3,4\} \mid \Omega\} \begin{pmatrix} 1 \\ 4 \end{pmatrix},
\end{aligned}$$

with the result

$$\begin{aligned}
Q\{\{1,2\} \mid \Omega\} &= \tfrac{1}{4} & Q\{\{3,4\} \mid \Omega\} &= \tfrac{3}{4} \\
Q\{\{1\} \mid \{1,2\}\} &= \tfrac{2}{3} & Q\{\{2\} \mid \{1,2\}\} &= \tfrac{1}{3} \\
Q\{\{3\} \mid \{3,4\}\} &= \tfrac{1}{3} & Q\{\{4\} \mid \{3,4\}\} &= \tfrac{2}{3}.
\end{aligned}$$

The unconditional Q-probabilities are the ones given above. Since all the conditional probabilities are uniquely determined, the unconditional probabilities are uniquely determined.

Market completeness can also be demonstrated by showing how to generate a given cash flow process. As already mentioned, it is sufficient to generate the cash flow processes in (5.45).

Consider first $\hat{t} = 1$ and let $A = \{1,2\} \in f_1$ and $B = \{3,4\} \in f_1$. Then the cash flow process $(1_A, 0)$ is attained by the following trading strategy:

$$\theta_1^1 + 8\theta_1^2 = 1 \wedge \theta_1^1 + 4\theta_1^2 = 0 \quad \Rightarrow \quad \theta_1^1 = -1, \ \theta_1^2 = \tfrac{1}{4}, \ \theta_1 s_0 = 0.25, \tag{5.47}$$

whereas the cash flow process $(1_B, 0)$ is attained by the following trading strategy:

$$\theta_1^1 + 8\theta_1^2 = 0 \wedge \theta_1^1 + 4\theta_1^2 = 1 \quad \Rightarrow \quad \theta_1^1 = 2, \ \theta_1^2 = -\tfrac{1}{4}, \ \theta_1 s_0 = 0.75. \tag{5.48}$$

Consider next $\hat{t} = 2$. Then

- the cash flow process $(0, 1_{\{1\}})$ is attained by the following trading strategy:

$$\begin{aligned} &\theta_2^1 + 9\theta_2^2 = 1 \wedge \theta_2^1 + 6\theta_2^2 = 0 \quad \Rightarrow \quad \theta_2^1 = -2, \ \theta_2^2 = \tfrac{1}{3}, \ \theta_2 s_1 = \tfrac{2}{3}, \\ &\theta_1^1 + 8\theta_1^2 = \tfrac{2}{3} \wedge \theta_1^1 + 4\theta_1^2 = 0 \quad \Rightarrow \quad \theta_1^1 = -\tfrac{2}{3}, \ \theta_1^2 = \tfrac{1}{6}, \ \theta_1 s_0 = \tfrac{1}{6}. \end{aligned} \tag{5.49}$$

- the cash flow process $(0, 1_{\{2\}})$ is attained by the following trading strategy:

$$\begin{aligned} &\theta_2^1 + 9\theta_2^2 = 0 \wedge \theta_2^1 + 6\theta_2^2 = 1 \quad \Rightarrow \quad \theta_2^1 = 3, \ \theta_2^2 = -\tfrac{1}{3}, \ \theta_2 s_1 = \tfrac{1}{3}, \\ &\theta_1^1 + 8\theta_1^2 = \tfrac{1}{3} \wedge \theta_1^1 + 4\theta_1^2 = 0 \quad \Rightarrow \quad \theta_1^1 = -\tfrac{1}{3}, \ \theta_1^2 = \tfrac{1}{12}, \ \theta_1 s_0 = \tfrac{1}{12}. \end{aligned} \tag{5.50}$$

- the cash flow process $(0, 1_{\{3\}})$ is attained by the following trading strategy:

$$\begin{aligned} &\theta_2^1 + 6\theta_2^2 = 1 \wedge \theta_2^1 + 3\theta_2^2 = 0 \quad \Rightarrow \quad \theta_2^1 = -1, \ \theta_2^2 = \tfrac{1}{3}, \ \theta_2 s_1 = \tfrac{1}{3}, \\ &\theta_1^1 + 8\theta_1^2 = 0 \wedge \theta_1^1 + 4\theta_1^2 = \tfrac{1}{3} \quad \Rightarrow \quad \theta_1^1 = \tfrac{2}{3}, \ \theta_1^2 = -\tfrac{1}{12}, \ \theta_1 s_0 = \tfrac{1}{4}. \end{aligned} \tag{5.51}$$

- the cash flow process $(0, 1_{\{4\}})$ is attained by the following trading strategy:

$$\begin{aligned} &\theta_2^1 + 6\theta_2^2 = 0 \wedge \theta_2^1 + 3\theta_2^2 = 1 \quad \Rightarrow \quad \theta_2^1 = 2, \ \theta_2^2 = -\tfrac{1}{3}, \ \theta_2 s_1 = \tfrac{2}{3}, \\ &\theta_1^1 + 8\theta_1^2 = 0 \wedge \theta_1^1 + 4\theta_1^2 = \tfrac{2}{3} \quad \Rightarrow \quad \theta_1^1 = \tfrac{4}{3}, \ \theta_1^2 = -\tfrac{1}{6}, \ \theta_1 s_0 = \tfrac{1}{2}. \end{aligned} \tag{5.52}$$

By construction these four trading strategies are self-financing.

Example 4 Consider the following modification of the price processes given in Example 3:

$$
\begin{array}{ccccc}
 & & & & \begin{pmatrix}1\\9\end{pmatrix} \\
 & & \begin{pmatrix}1\\8\end{pmatrix} & \begin{matrix}\nearrow\\ \searrow\end{matrix} & \\
 & \nearrow & & & \begin{pmatrix}1\\6\end{pmatrix} \\
\begin{pmatrix}1\\5\end{pmatrix} & & & & \\
 & \searrow & & & \begin{pmatrix}1\\6\end{pmatrix} \\
 & & \begin{pmatrix}1\\4\end{pmatrix} & \begin{matrix}\nearrow\\ \searrow\end{matrix} & \\
 & & & & \begin{pmatrix}1\\4\end{pmatrix}
\end{array}
$$

In this example, the market is complete. However, an arbitrage opportunity exists. Consider, for example, the trading strategy:

$$\theta_1 = 0, \qquad \theta_2(\{1,2\}) = 0, \qquad \theta_2(\{3,4\}) = \begin{pmatrix} -4 \\ 1 \end{pmatrix}.$$

The cash flow process from this trading strategy is:

$$\delta_0 = \delta_1(\theta)(\{1,2\}) = \delta_1(\theta)(\{3,4\}) = 0$$

$$\delta_2(\theta)(\{1\}) = \delta_2(\theta)(\{2\}) = \delta_2(\theta)(\{4\}) = 0 \qquad \delta_2(\theta)(\{3\}) = 2.$$

Other examples can be constructed. But it is not in general possible to determine in advance, where a strict inequality – as required by the definition of an arbitrage opportunity – occurs. Given the presence of an asset with a strictly positive price process, any gain from an arbitrage opportunity can be carried forward and liquidated at a later date. On the other hand it is generally impossible to liquidate a future arbitrage opportunity before it actually occurs.

Example 5 Consider yet another modification of these examples:

$$
\begin{pmatrix} 1 \\ 5 \end{pmatrix} \nearrow \begin{pmatrix} 1 \\ 8 \end{pmatrix}
\begin{array}{l} \nearrow \\ \rightarrow \\ \searrow \end{array}
\begin{array}{l} \begin{pmatrix} 1 \\ 9 \end{pmatrix} \\ \begin{pmatrix} 1 \\ 8 \end{pmatrix} \\ \begin{pmatrix} 1 \\ 6 \end{pmatrix} \end{array}
$$

$$
\searrow \begin{pmatrix} 1 \\ 4 \end{pmatrix}
\begin{array}{l} \nearrow \\ \searrow \end{array}
\begin{array}{l} \begin{pmatrix} 1 \\ 6 \end{pmatrix} \\ \begin{pmatrix} 1 \\ 3 \end{pmatrix} \end{array}
$$

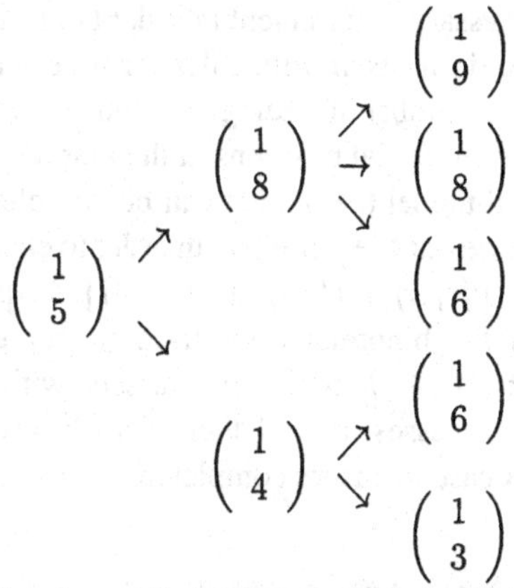

In this example, the market is not complete. However, no arbitrage opportunities exist. Enumerating again the paths 'downwards', the incompleteness arises because at atom $\{1, 2, 3\} \in \mathcal{F}_1$ we have a splitting index 3, but $\mathrm{rank}(S(1, \{1, 2, 3\})) = 2$. Any of the following measures

ω	1	2	3	4	5
$Q(\omega)$	$2q$	$\frac{1}{4} - 3q$	q	$\frac{1}{4}$	$\frac{1}{2}$

with $0 < q < \frac{1}{12}$ are martingale measures. In order to illustrate, let us fix the following two martingale measures, corresponding to $q = \frac{1}{24}$ and $q^* = \frac{1}{16}$, respectively, together with the Radon–Nikodym derivative $\rho_X \equiv \frac{dQ^*}{dQ}$:

ω	1	2	3	4	5
$Q\{\omega\}$	$\frac{1}{12}$	$\frac{1}{8}$	$\frac{1}{24}$	$\frac{1}{4}$	$\frac{1}{2}$
$Q^*\{\omega\}$	$\frac{1}{8}$	$\frac{1}{16}$	$\frac{1}{16}$	$\frac{1}{4}$	$\frac{1}{2}$
$\rho_X(\omega)$	$\frac{3}{2}$	$\frac{1}{2}$	$\frac{3}{2}$	1	1

The vector X is in $\mathcal{R}^7$. It is uniquely determined up to a scalar, and it can easily be verified that with respect to Q the vector

$$
X_1 = \begin{pmatrix} 0 \\ 0 \end{pmatrix} \qquad X_2 = \begin{pmatrix} 1 \\ -1 \\ 1 \\ 0 \\ 0 \end{pmatrix},
$$

does the job.

It is worth commenting on the relation between the splitting function for the partition, the number of trading dates, and the number of traded assets $N + 1$.

In the 'two-period' model at the beginning of this paper we needed $N+1 \geq S$ as a necessary requirement in order to ensure completeness. This had nothing to do in itself with calendar time – the only relevant consideration was the number of trading dates on which individuals were allowed to trade their financial positions in the market.

In the multiperiod model the market can be complete with far fewer traded assets. The necessary requirement in order to ensure completeness is $N+1 \geq \max\{\nu(t,\omega) \mid (t,\omega) \in \mathcal{T} \times \Omega\}$. A particularly well-known example is the binomial case, used in our examples, where $\nu(t,\omega) = 2 \;\; \forall (t,\omega) \in \mathcal{T} \times \Omega$. Two assets with locally linearly independent price processes, one of them usually a one-period bond, are sufficient in this case to ensure completeness.

5.9 Complete market and martingale basis

Definition 8 Let $(\Omega, \mathcal{F}, P)$ and the filtration $\{\mathcal{F}_t\}$ be given. (We drop the distinction between calendar time and trading dates here for notational convenience.) Let

$$M^1, M^2, \ldots, M^p$$

be martingales relative to P and the filtration $\{\mathcal{F}_t\}$. These martingales are said to be linearly independent if

$$\begin{aligned} \sum_{i=1}^{p} \eta^i \cdot M^i &= 0, \quad \eta^i \text{ predictable } \forall i = 1, \ldots, p \\ &\Rightarrow \quad \eta^i = 0 \;\; \forall\, i = 1, \ldots, p. \end{aligned}$$

The martingales are said to form a basis for all $(P, \{\mathcal{F}_t\})$-martingales if

1. for any $(P,\{\mathcal{F}_t\})$-martingale X there exist predictable processes $\eta^1, \eta^2, \ldots, \eta^p$ so that

$$X_t = X_0 + \sum_{\tau=1}^{t} \sum_{i=1}^{p} \eta^i_\tau \cdot (M^i_\tau - M^i_{\tau-1});$$

2. no proper subset of these p martingales satisfies this property.

Theorem 9 Assume $\nu(t,\omega) = p > 1 \;\; \forall (t,\omega) \in \mathcal{T} \times \Omega$. Then $p-1$ is the dimension of any martingale basis.

Proof Observe that any $\{f_u\}_{u \in \mathcal{T}}$-adapted stochastic process x_u, $0 \leq u \leq t$, with values in $\{1, 2, \ldots, p\}$, and for which $x_u : \Omega \to \{1, 2, \ldots, p\}$

is a surjective mapping for $0 < u \leq T$, generates the partition associated with the filtration $\{\mathcal{F}_t\}$, and hence also generates the filtration.

Let M be any $(P, \{\mathcal{F}_t\})$-martingale. M can thus be expressed as

$$M_t = g_t(x_0, x_1, \ldots, x_t) = \sum_{k=1}^{p} g_t(x_0, x_1, \ldots, x_{t-1}, k) \cdot 1_{\{x_t=k\}}$$

for suitable $\mathcal{F}_t$-measurable Borel functions g_t. Consequently

$$\begin{aligned} M_{t-1} &= E\left[M_t \mid \mathcal{F}_{t-1}\right] \\ &= E\left[\sum_{k=1}^{p} g_t(x_0, x_1, \ldots, x_{t-1}, k) \cdot 1_{\{x_t=k\}} \mid \mathcal{F}_{t-1}\right] \\ &= \sum_{k=1}^{p} g_t(x_0, x_1, \ldots, x_{t-1}, k) \cdot E\left[1_{\{x_t=k\}} \mid \mathcal{F}_{t-1}\right], \end{aligned}$$

since $g_t(x_0, x_1, \ldots, x_{t-1}, k)$ is $\mathcal{F}_{t-1}$-measurable. By subtraction we obtain

$$M_t - M_{t-1} = \sum_{k=1}^{p} g_t(x_0, x_1, \ldots, x_{t-1}, k) \cdot \left[1_{\{x_t=k\}} - E\left[1_{\{x_t=k\}} \mid \mathcal{F}_{t-1}\right]\right]$$

Since

$$\sum_{k=1}^{p} \left[1_{\{x_t=k\}} - E\left[1_{\{x_t=k\}} \mid \mathcal{F}_{t-1}\right]\right] = 0,$$

it is also the case that

$$\begin{aligned} M_t - M_{t-1} = \sum_{k=1}^{p-1} &\left[g_t(x_0, x_1, \ldots, x_{t-1}, k) - g_t(x_0, x_1, \ldots, x_{t-1}, p)\right] \cdot \\ &\left[1_{\{x_t=k\}} - E\left[1_{\{x_t=k\}} \mid \mathcal{F}_{t-1}\right]\right]. \end{aligned}$$

This is a representation as claimed, since

- $g_t(x_0, x_1, \ldots, x_{t-1}, k) - g_t(x_0, x_1, \ldots, x_{t-1}, p)$ is a predictable process and
- $1_{\{x_t=k\}} - E\left[1_{\{x_t=k\}} \mid \mathcal{F}_{t-1}\right]$ is the martingale increment.

It is worth noting that the martingale increments

$$N_t^k - N_{t-1}^k \equiv 1_{\{x_t=k\}} - E\left[1_{\{x_t=k\}} \mid \mathcal{F}_{t-1}\right] \quad (5.53)$$

$$N_t^k = \sum_{j=1}^{t} \left[1_{\{x_j=k\}} - E\left[1_{\{x_j=k\}} \mid \mathcal{F}_{j-1}\right]\right] \quad (5.54)$$

are independent of the particular martingale M. It remains to be shown that the basis found is linearly independent.

Assume that

$$\sum_{k=1}^{p-1} \eta_t^k N_t^k = 0 \quad \text{and that at least one } \eta_t^h \neq 0.$$

Then

$$\sum_{k=1}^{p-1} \eta_t^k \sum_{j=1}^{t} 1_{\{x_j=k\}} = \sum_{k=1}^{p-1} \eta_t^k \sum_{j=1}^{t} E\left[1_{\{x_j=k\}} \mid \mathcal{F}_{j-1}\right]. \tag{5.55}$$

The right-hand side of (5.55) is $\mathcal{F}_{t-1}$-measurable. So is η_t^k by the assumption of predictability. The indicator functions $1_{\{x_j=k\}}$, $j = 1, 2, \ldots, t-1$, are $\mathcal{F}_{t-1}$-measurable. Hence, (5.55) can be reduced to

$$\sum_{k=1}^{p-1} \eta_t^k 1_{\{x_t=k\}} = \sum_{k=1}^{p-1} \eta_t^k E\left[1_{\{x_t=k\}} \mid \mathcal{F}_{t-1}\right]. \tag{5.56}$$

The sum on the left-hand side picks exactly one element k, where $1_{\{x_t=k\}} = 1$ and all other terms are zero. By assumption there is at least one atom in $\mathcal{F}_{t-1}$ such that $\eta_t^h \neq 0$. And this atom splits into $p > 1$ new atoms in $\mathcal{F}_t$. Hence, the left-hand side of (5.55) can take on two different values, 0 and $\eta_t^h \neq 0$, at the same atom in $\mathcal{F}_{t-1}$, which contradicts the $\mathcal{F}_{t-1}$-measurability.

Example 6 Consider the standard binomial model with constant interest rate r. For this model $p = 2$, and the discounted price process evolves according to the following tree:

$$\begin{array}{ccccc}
 & & & & \begin{pmatrix} 1 \\ u^2 S_{t-1}/(1+r)^2 \end{pmatrix} \\
 & & \begin{pmatrix} 1 \\ uS_{t-1}/(1+r) \end{pmatrix} & \begin{array}{c}\nearrow \\ \searrow\end{array} & \\
 & \nearrow & & & \begin{pmatrix} 1 \\ udS_{t-1}/(1+r)^2 \end{pmatrix} \\
\begin{pmatrix} 1 \\ S_{t-1} \end{pmatrix} & & & & \\
 & \searrow & & & \begin{pmatrix} 1 \\ duS_{t-1}/(1+r)^2 \end{pmatrix} \\
 & & \begin{pmatrix} 1 \\ dS_{t-1}/(1+r) \end{pmatrix} & \begin{array}{c}\nearrow \\ \searrow\end{array} & \\
 & & & & \begin{pmatrix} 1 \\ d^2 S_{t-1}/(1+r)^2 \end{pmatrix}
\end{array}$$

with $u > 1 + r > d$.

For this model, the predictable process becomes

$$g(x_0, x_1, \ldots, x_{t-1}, 1) = \frac{u}{1+r} \cdot S_{t-1},$$

$$g(x_0, x_1, \ldots, x_{t-1}, 2) = \frac{d}{1+r} \cdot S_{t-1}.$$

It is well known that the martingale measure has probabilities $\theta = \frac{1+r-d}{u-d}$ of an 'up' move and $1 - \theta = \frac{u-(1+r)}{u-d}$ of a 'down' move. Associating $k = 1$ with an 'up' move and $k = 2$ with a 'down' move we have the martingale increment

$$\begin{pmatrix} 1_{\{x_t=1\}} - E\left[1_{\{x_t=1\}} \mid \mathcal{F}_{t-1}\right] \\ 1_{\{x_t=1\}} - E\left[1_{\{x_t=1\}} \mid \mathcal{F}_{t-1}\right] \end{pmatrix} = \begin{pmatrix} 1-\theta \\ -\theta \end{pmatrix}.$$

This standard binomial model has served as the premier example of derivative asset pricing with a stock as underlying asset since it first appeared (see Cox, Ross and Rubinstein, 1979). It is easily generalized to include foreign exchange as underlying asset. However, introducing interest rate uncertainty in this framework quickly leads to complications, the exposition of which goes beyond the scope of this paper. For an in-depth treatment of pricing of bonds and interest rate derivatives in the binomial setting, see Jensen and Nielsen (1992).

We finish our exposition of discrete-time models with a characterization of the attainable cash flow processes. Not surprisingly, the attainable cash flow processes in an arbitrage-free market are precisely the cash flow processes arising from applying trading strategies to the $N + 1$ basic assets.

Theorem 10 Assume that no arbitrage strategies exist. Let Q be an equivalent martingale measure with s^0 as numéraire. The cash flow process $\{\delta_t\}$ is attainable if and only if there exists a $(N + 1)$-dimensional, predictable process $\{\eta_t\}$, so that

$$E^Q\left[\sum_{v=1}^{T} \tilde{\delta}_v \mid \mathcal{F}_t\right] = E^Q\left[\sum_{v=1}^{T} \tilde{\delta}_v\right] + \sum_{v=1}^{t}\sum_{i=0}^{N} \eta_v^i \cdot \left(\tilde{s}_v^i - \tilde{s}_{v-1}^i\right). \quad (5.57)$$

Proof Assume that $(\delta_1, \delta_2, \ldots, \delta_T)$ is attainable. Observe that a given cash flow process is attainable if and only if there exists a trading strategy θ_v, $v = 1, 2, \ldots, T + 1$, with $\theta_{T+1} = 0$, so that

$$\sum_{v=1}^{t} \tilde{\delta}_v + \theta_{t+1} \cdot \tilde{s}_t = \theta_1 \cdot \tilde{s}_0 + \sum_{v=1}^{t}\sum_{i=0}^{N} \theta_v^i \cdot \left(\tilde{s}_v^i - \tilde{s}_{v-1}^i\right) \ \ \forall t = 1, 2, \ldots, T. \quad (5.58)$$

As θ_v^i is predictable and $\left(\tilde{s}_v^i - \tilde{s}_{v-1}^i\right)$ is a martingale increment, the last sum in (5.58) is a Q-martingale. Applying (5.58) with $t = T$ and conditioning on $\mathcal{F}_t$, we obtain the last sum in (5.57) with θ_v^i being the predictable process. The first term follows from taking the unconditional mean in (5.58) with $t = T$.

To prove the reverse, it is necessary to construct a trading strategy which – whenever (5.57) is given – generates the cash flow process. Consider, for example, the trading strategy

$$\begin{aligned}
\theta_t^i &= \eta_t^i \quad \text{for } i = 1, 2, \ldots, N \text{ and } t = 1, 2, \ldots, T, \\
\theta_1^0 &= E^Q\left[\sum_{v=1}^{T} \tilde{\delta}_v\right] - \sum_{i=1}^{N} \theta_1^i \cdot \tilde{s}_0^i, \\
\theta_{t+1}^0 &= E^Q\left[\sum_{v=1}^{T} \tilde{\delta}_v \mid \mathcal{F}_t\right] - \sum_{v=1}^{t} \tilde{\delta}_v - \sum_{i=1}^{N} \theta_{t+1}^i \cdot \tilde{s}_t^i \\
&\quad \text{for } t = 1, \ldots, T.
\end{aligned}$$

Substituting the two expectation terms from the prescribed trading strategy into (5.57), we obtain

$$\sum_{v=1}^{t} \tilde{\delta}_v = \theta_1 \cdot \tilde{s}_0 - \theta_{t+1} \cdot \tilde{s}_t + \sum_{v=1}^{t} \sum_{i=0}^{N} \eta_v^i \cdot \left(\tilde{s}_v^i - \tilde{s}_{v-1}^i\right). \qquad (5.59)$$

Observe that $\tilde{s}_v^0 \equiv 1 \ \forall v = 0, 1, \ldots, T$. Hence, the predictable processes η_v^i can be replaced by the prescribed trading strategy θ_v^i to give the desired result.

Corollary 3 Assume that no arbitrage strategies exist, and that Q is an equivalent martingale measure. The market is complete if and only if any Q–martingale X_t has a representation in terms of a martingale transform

$$X_t = X_0 + \sum_{v=1}^{t} \sum_{i=0}^{N} \eta_v^i \cdot \left(\tilde{s}_v^i - \tilde{s}_{v-1}^i\right), \qquad (5.60)$$

with η_t being a predictable process.

Proof Let X_t be a Q-martingale and consider the cash flow process

$$\tilde{\delta}_t = 0 \quad \text{for } t = 1, 2, \ldots, T-1 \qquad \tilde{\delta}_T = X_T.$$

Since X_t is a Q-martingale it follows that

$$X_t = E^Q\left[\sum_{v=1}^{T} \tilde{\delta}_v \mid \mathcal{F}_t\right].$$

Since $\tilde{\delta}_v$ is attainable – due to the complete market assumption – we obtain the desired representation from Theorem 10.

For any cash flow process $\tilde{\delta}_v$ we have that

$$X_t = E^Q \left[\sum_{v=1}^{T} \tilde{\delta}_v \mid \mathcal{F}_t \right]$$

is a Q-martingale. Hence it has a martingale transform representation. By Theorem 10 it follows that the cash flow process is attainable, and the corollary is proved.

To summarize: within the setting of discrete-time processes on a finite state space we have obtained several conclusions – based on the simple 'no arbitrage' condition – concerning the pricing of derivative assets .

There is an equivalence between the following statements:

- no arbitrage opportunities
- linear pricing rule
- the existence of at least one equivalent martingale measure.

The linear pricing rule has an intuitive economic interpretation in markets that function as well as the financial markets. And the equivalent martingale measure gives access to the well-developed theory on martingales; as stated in Harrison and Pliska (1981), 'we have started to feel that all the standard problems studied in martingale theory and all the major results have interpretations and applications in our setting.' Furthermore, we showed that in some situations all contingent claims can be priced uniquely. This is the case if the following equivalent requirements are met:

- the market is complete
- the equivalent martingale measure Q is unique
- the relative price processes $\tilde{s}^1, \tilde{s}^2, \ldots, \tilde{s}^N$ generate all Q-martingales.

5.10 Continuous-time models: basic assumptions

In the discrete-time models developed in earlier sections, the mathematical tools required proved to be quite modest. Despite this, no genuine new qualitative insight is obtained by moving to continuous time. The menus of price processes, trading strategies, etc., are markedly increased, but simultaneously one incurs costs of dealing with a number of regularity conditions. Such regularity conditions are of varying complexity, but generally do not add much economic insight.

In continuous-time models we have $\mathcal{T} = [0, T]$. We restrict ourselves to price processes, represented as Itô processes of the form

$$S_t^i = S_0^i + \int_0^t \mu_s^i ds + \int_0^t \sigma_s^{i1} dB_s^1 + \ldots + \int_0^t \sigma_s^{id} dB_s^d \quad i = 0, 1, \ldots, N, \tag{5.61}$$

where

- B_s^j, $j = 1, 2, \ldots, d$ denote independent standard Brownian motions;
- μ_s^i and σ_s^{ij} are adapted processes;
- $\boldsymbol{\mu}_s$ is the $(N+1)$-dimensional vector $(\mu_s^0, \mu_s^1, \ldots, \mu_s^N)$;
- $\boldsymbol{\sigma}_s$ is a $((N+1) \times d)$-dimensional matrix, with $\boldsymbol{\sigma}_s \cdot \boldsymbol{\sigma}_s'$ being the variance-covariance matrix of the price process dynamics;
- the integrals are well defined and the stochastic integrals are $\mathcal{F}_t$-martingales. Sufficient conditions, which are adopted here, are:

$$E^P\left[\int_0^T (\mu_s^i)^2 ds\right] < \infty \quad \text{and} \quad E^P\left[\int_0^T \| \sigma_s^i \|^2 ds\right] < \infty.$$

The filtration is taken to be the filtration generated by the traded assets

$$\mathcal{F}_\tau = \sigma\{s_t^0, s_t^1, \ldots, s_t^N\} \quad 0 \leq t \leq \tau,$$

so that any price process is automatically an adapted process. We need to add as a technical assumption that $\mathcal{F}_0$ is complete. Furthermore, equalities and inequalities will be understood to be 'P-a.s.' in what follows.

Trading takes place continuously. More appropriately, one could say that it is possible for an investor to trade continuously if so desired. Because of this, some regularity conditions must be imposed in order to eliminate so-called 'doubling strategies' (the continuous-time analogue of St Petersburg paradox strategies) as well as so-called 'suicide strategies'.

Definition 9 A trading strategy is a predictable, stochastic process $\boldsymbol{\theta}_t$. The associated cash flow process $\delta_t(\boldsymbol{\theta})$ is the process determined by

$$\begin{aligned} \int_0^t \delta_s(\boldsymbol{\theta}) ds &= -\boldsymbol{\theta}_t \cdot \boldsymbol{S}_t + \boldsymbol{\theta}_0 \cdot \boldsymbol{S}_0 + \int_0^t \boldsymbol{\theta}_s \cdot d\boldsymbol{S}_s \\ &= \int_0^t \boldsymbol{\theta}_s \cdot d\boldsymbol{S}_s - \int_0^t d(\boldsymbol{\theta}_s \cdot \boldsymbol{S}_s) \end{aligned} \tag{5.62}$$

$$\begin{aligned} \Leftrightarrow \quad & \boldsymbol{\theta}_t \cdot \boldsymbol{S}_t \\ = \quad & \boldsymbol{\theta}_0 \cdot \boldsymbol{S}_0 - \int_0^t \delta_s(\boldsymbol{\theta}) ds + \int_0^t \boldsymbol{\theta}_s \cdot d\boldsymbol{S}_s. \end{aligned} \tag{5.63}$$

Definition 10 The adapted process $V_t(\boldsymbol{\theta}) \equiv \boldsymbol{\theta}_t \cdot \boldsymbol{S}_t$ is called the **value process** associated with the trading strategy $\boldsymbol{\theta}_t$. It denotes the value at time t of the portfolio of the $N + 1$ assets given by the trading strategy $\boldsymbol{\theta}_t$.

By this definition of the cash flow process it is automatically assumed that the sample functions on the right-hand side of (5.62) are a.s. absolutely continuous. This assumption is too strong for many applications, and it can easily be generalized to the case where the right-hand side of (5.62) is a.s. of bounded variation. However, we will not go into any discussion along these lines, since most arguments in the rest of this paper anyway rely upon the right-hand side of (5.62) being identically zero.

Definition 11 A self-financing trading strategy is a trading strategy $\boldsymbol{\theta}_t$ satisfying

$$\boldsymbol{\theta}_t \cdot \boldsymbol{S}_t = \boldsymbol{\theta}_0 \cdot \boldsymbol{S}_0 + \int_0^t \boldsymbol{\theta}_s \cdot d\boldsymbol{S}_s \qquad \forall t \in [0, T]. \tag{5.64}$$

This is equivalent to $\delta_t(\boldsymbol{\theta}) = 0 \, \forall t \in [0, T]$.

Any trading strategy that is allowed must preserve some of the analytical properties required by the price processes of individual assets. The following are natural requirements, given the requirements on asset price processes themselves:

$$E^P\left[\int_0^T (\boldsymbol{\theta}_t \cdot \boldsymbol{\mu}_t)^2 dt\right] < \infty, \tag{5.65}$$

$$E^P\left[\int_0^T \| \boldsymbol{\theta}_t \cdot \boldsymbol{\sigma}_t \|^2 \, dt\right] < \infty. \tag{5.66}$$

The interpretation of the cash flow process is entirely analogous to the cash flow process in discrete time. Given that we have restricted ourselves to non-dividend-paying assets, the portfolio value $\boldsymbol{\theta}_t \cdot \boldsymbol{S}_t$ can only arise from (i) an initial payment $\boldsymbol{\theta}_0 \cdot \boldsymbol{S}_0$, (ii) (net) payments $\delta_s(\boldsymbol{\theta})$ from the investor into the portfolio, and (iii) capital gains $\boldsymbol{\theta}_s \cdot d\boldsymbol{S}_s$. The integrability conditions in (5.65) and (5.66) are sufficient conditions to ensure that $\int_0^t \boldsymbol{\theta}_s \cdot d\boldsymbol{S}_s$ is a finite-variance process and that $\int_0^t \boldsymbol{\theta}_s \cdot \boldsymbol{\sigma}_s \cdot d\boldsymbol{B}_s$ is a $\mathcal{F}_t$-martingale.

Definition 12 Let $Z \in L^2(\Omega, \mathcal{F}_T, P)$. Z is said to be attainable and to be financed by the self-financing trading strategy $\boldsymbol{\theta}_t$ if a.s.:

$$V_t(\boldsymbol{\theta}) \equiv \boldsymbol{\theta}_t \cdot \boldsymbol{S}_t = \boldsymbol{\theta}_0 \cdot \boldsymbol{S}_0 + \int_0^t \boldsymbol{\theta}_s \cdot d\boldsymbol{S}_s, \quad \boldsymbol{\theta}_T \cdot \boldsymbol{S}_T = Z. \quad (5.67)$$

5.11 Relative price processes and change of numéraire

Referring to the definition of a self-financing trading strategy, it is clear that the value process $V_t(\boldsymbol{\theta})$ will be a martingale if and only if $\int \boldsymbol{\theta}_s d\boldsymbol{S}_s$ is a martingale. This will only be the case whenever $\boldsymbol{\mu}_t : [0,T] \times \Omega \rightarrow \mathcal{R}^{N+1}$ is zero. Analogous to the derivations in discrete time, pricing by 'no arbitrage' involves (i) a change of numéraire and (ii) a change of measure. In this section we will focus on changing the numéraire.

Assume that asset no. 0 has a positive price process S_t^0 a.s. We denote the asset price processes, when expressed in terms of this numéraire, as

$$\tilde{S}_t \equiv \frac{\boldsymbol{S}_t}{S_t^0} = \left(1, \frac{S_t^1}{S_t^0}, \frac{S_t^2}{S_t^0}, \ldots, \frac{S_t^N}{S_t^0}\right). \quad (5.68)$$

Without further assumptions, a routine application of Itô's lemma leads to

$$\begin{aligned} \tilde{S}_t^i &= \tilde{S}_0^i + \int_0^t \tilde{\mu}_s^i ds + \sum_{j=1}^d \int_0^t \tilde{\sigma}^{ij} dB_s^j \\ &\equiv \tilde{S}_0^i + \int_0^t \left(\frac{\mu_s^i}{S_s^0} - \frac{S_s^i}{(S_s^0)^2}\mu_s^0 + \frac{S_s^i}{(S_s^0)^3}\sum_{j=1}^d \left(\sigma_s^{0j}\right)^2 \right. \\ &\quad - \left. \frac{1}{(S_s^0)^2}\sum_{j=1}^d \sigma_s^{0j}\sigma_s^{ij} \right) ds + \\ &\quad \sum_{j=1}^d \int_0^t \left(\frac{\sigma_s^{ij}}{S_s^0} - \frac{S_s^i \sigma_s^{0j}}{(S_s^0)^2} \right) dB_s^j. \end{aligned} \quad (5.69)$$

It is a rather demanding task to ensure simultaneously that the stochastic integrals are well defined as martingales and that the feasible trading strategies are unchanged or at least identifiable. Consider

$$\int_0^t \boldsymbol{\theta}_s d\left(\tilde{S}^0, \tilde{\boldsymbol{S}}\right)_s = \int_0^t \boldsymbol{\theta}_s \left(0, \tilde{\boldsymbol{\mu}}_s\right) ds + \int_0^t \boldsymbol{\theta}_s \begin{pmatrix} 0 \\ \tilde{\boldsymbol{\sigma}}_s \end{pmatrix} d\boldsymbol{B}_s. \quad (5.70)$$

Let $\overline{\boldsymbol{\theta}}_t \equiv (\theta_t^1, \theta_t^2, \ldots, \theta_t^N)$ denote the trading strategy with the zeroth component omitted. In order to ensure that a feasible trading strategy for the price processes $\boldsymbol{S}_t$ remains a feasible trading strategy for the price processes $\tilde{\boldsymbol{S}}_t$ we require

$$E^P\left[\int_0^T \left(\overline{\boldsymbol{\theta}}_t \cdot \tilde{\boldsymbol{\mu}}_t\right)^2 dt\right] \quad < \quad \infty, \tag{5.71}$$

$$E^P\left[\int_0^T \| \overline{\boldsymbol{\theta}}_t \cdot \tilde{\boldsymbol{\sigma}}_t \|^2 dt\right] \quad < \quad \infty. \tag{5.72}$$

This leads to the following definition.

Definition 13 A **feasible numéraire** is a strictly positive Itô process fulfilling (5.71) and (5.72) for all trading strategies $\boldsymbol{\theta}_s$ satisfying (5.65) and (5.66).

A natural question is whether such numéraires actually exist. This is at least the case when the ubiquitous 'bank account' is used as numéraire. The value of the bank account:

$$S_t^0 = \exp\left(\int_0^t r_s \, ds\right) \tag{5.73}$$

is a price process of bounded variation, i.e. $\sigma_S^{0j} = 0$. And provided that the short-term interest rate r_s is a bounded process, it is easy to see that (5.65)–(5.66) and (5.71)–(5.72) are equivalent.

Theorem 11 Assume that S_t^0 is a feasible numéraire. Then $\boldsymbol{\theta}_t$ is a self-financing trading strategy with respect to $\tilde{\boldsymbol{S}}_t$ if $\boldsymbol{\theta}_t$ is self-financing trading strategy with respect to $\boldsymbol{S}_t$.

Proof First of all, self-financing trading strategies can be characterized by the property that the cash-flow process is identically zero: $\delta_t(\boldsymbol{\theta}) = 0$. The cash flow process will remain zero independent of any particular choice of numéraire.

A more complete proof goes as follows. Assume that $\boldsymbol{\theta}_t$ is self-financing in the given numéraire. Then

$$V_t(\boldsymbol{\theta}) = \boldsymbol{\theta}_t \cdot \boldsymbol{S}_t \quad \Rightarrow \quad \tilde{V}_t(\boldsymbol{\theta}) \equiv \frac{V_t(\boldsymbol{\theta})}{S_t^0} = \boldsymbol{\theta}_t \cdot \frac{\boldsymbol{S}_t}{S_t^0} \equiv \boldsymbol{\theta}_t \cdot \tilde{\boldsymbol{S}}_t$$

$$\begin{aligned} dV_t(\boldsymbol{\theta}) = \boldsymbol{\theta}_t \cdot d\boldsymbol{S}_t \quad \Rightarrow \quad d\tilde{V}_t(\boldsymbol{\theta}) &= dV_t(\boldsymbol{\theta})\frac{1}{S_t^0} + V_t(\boldsymbol{\theta})d\left(\frac{1}{S_t^0}\right) \\ &+ \quad dV_t(\boldsymbol{\theta})d\left(\frac{1}{S_t^0}\right) \end{aligned}$$

$$= \quad \theta_t \cdot \left[\frac{1}{S_t^0} dS_t + S_t d\left(\frac{1}{S_t^0}\right) + dS_t d\left(\frac{1}{S_t^0}\right)\right]$$
$$= \quad \theta_t \cdot d\tilde{S}_t. \tag{5.74}$$

Hence, θ_t is self-financing in the numéraire S_t^0.

Reversing the argument requires that $\frac{1}{S_t^0}$ is a feasible numéraire. This is the case with the bank account, but in general it requires some boundedness conditions. We will not pursue the details here.

Definition 14 An **arbitrage opportunity** or an **arbitrage strategy** for the prices S_t is a self-financing trading strategy θ_t with the properties:

$$\theta_0 \cdot S_0 \leq 0, \qquad \theta_T \cdot S_T \geq 0, \qquad P\{\theta_T \cdot S_T > 0\} > 0. \tag{5.75}$$

Theorem 12 Let S_t^0 be a feasible numéraire. Then there are no arbitrage strategies for the relative prices $\tilde{S}_t$ if there are no arbitrage strategies for the prices S_t.

Proof From Theorem 11 it follows that the property of being self-financing is invariant with respect to a change of numéraire. Furthermore, since S_t^0 is a strictly positive process,

$$\theta_0 \cdot S_0 \leq 0, \qquad \theta_T \cdot S_T \geq 0, \qquad P\{\theta_T \cdot S_T > 0\} > 0 \qquad \Leftrightarrow$$
$$\theta_0 \cdot \tilde{S}_0 \leq 0, \qquad \theta_T \cdot \tilde{S}_T \geq 0, \qquad P\{\theta_T \cdot \tilde{S}_T > 0\} > 0.$$

This theorem has some far-reaching consequences as to the pricing of assets. However, a change of measure is generally needed.

Definition 15 Q is an equivalent martingale measure for $\tilde{S}_t$ if

- $\tilde{S}_t$ is a Q-martingale
- $\frac{dQ}{dP} \in L^2(\Omega, \mathcal{F}_T, P)$.

In discrete time, the equivalence between the 'no arbitrage' condition and the existence of (at least one) equivalent martingale measure was established using modest technical tools. Theorem 13 is a 'one-way' theorem and is the 'easy part' of this in continuous time.

Theorem 13 Let S_t^0 be a feasible numéraire. If there exists an equivalent martingale measure, Q, for $\tilde{S}_t$, then there are no arbitrage opportunities.

Proof The standard Brownian motion under Q is denoted by $\tilde{B}_t$. By construction we have

$$d\tilde{S}_t = \tilde{\sigma}_t \cdot d\tilde{B}_t. \tag{5.76}$$

Hence

$$\int_0^T \theta_s \cdot d\tilde{S}_s = \int_0^T \theta_s \cdot \tilde{\sigma}_s \cdot d\tilde{B}_s \tag{5.77}$$

is a Q-martingale. We only need to verify the sufficient condition (this condition is known to be sufficient for the stochastic integral to be a martingale besides the previous one stated without the exponent 1/2)

$$E^Q\left[\left(\int_0^T \| \theta_s \cdot \tilde{\sigma}_s \|^2 \, ds\right)^{1/2}\right] < \infty. \tag{5.78}$$

Under the P-measure, the trading strategy θ_t is subject to the restriction

$$E^P\left[\int_0^T \| \theta_s \cdot \tilde{\sigma}_s \|^2 \, ds\right] < \infty,$$

and by assumption

$$E^P\left[\left(\frac{dQ}{dP}\right)^2\right] < \infty.$$

By Hölder's inequality we have that

$$E^P\left[\frac{dQ}{dP} \cdot \left(\int_0^T \| \theta_s \cdot \tilde{\sigma}_s \|^2 \, ds\right)^{1/2}\right] \leq$$

$$\left[E^P\left(\frac{dQ}{dP}\right)^2\right]^{1/2} \cdot \left[E^P\left(\int_0^T \| \theta_s \cdot \tilde{\sigma}_s \|^2 \, ds\right)\right]^{1/2} < \infty.$$

Hence, (5.78) is fulfilled, implying that

$$E^Q\left[\theta_T \cdot \tilde{S}_T\right] = \theta_0 \cdot \tilde{S}_0. \tag{5.79}$$

Absence of arbitrage follows immediately.

Nothing has been said so far about how an equivalent martingale measure can be found or even whether such a measure exists. As a special case one could try to look for a feasible numéraire such that the relative price processes become martingales under the original probability measure P. This is sometimes possible. An interesting – but seemingly overlooked – example of this approach is the methodology found in the pioneering article by Vasicek (1977).

It is difficult to give the integrability conditions (5.65) and (5.66) an economic interpretation. Hence, attempts have been made to limit the admissible trading strategies in interpretable ways. The integrability conditions (5.65) and (5.66) can be substituted by a non-negativity restriction $V_t(\boldsymbol{\theta}) \geq 0\, \forall t$. These two conditions are not entirely equivalent, unless all of $L^2(Q)$ is attainable by means of trading strategies fulfilling (5.65) and (5.66). In general, they are only identical 'up to closure' with the attainable part of $L^2(Q)$ under (5.65) and (5.66) being a subspace of the attainable part of $L^2(Q)$ under the non-negativity condition. However, in order to prove Theorem 13, both delimitations are equally useful.

As a central theorem for finding Radon–Nikodym derivatives in stochastic calculus we state – without proof – the following well known theorem.

Theorem 14 (Girsanov's Theorem) Let $\boldsymbol{\eta}_t \equiv (\eta_t^1, \eta_t^2, \ldots, \eta_t^d)$ be a vector of adapted processes satisfying

$$\int_0^T \| \boldsymbol{\eta}_t \|^2 \, dt < \infty.$$

Define $L(t)$ by

$$L(t) = \exp\left\{ - \int_0^t \boldsymbol{\eta}_s \cdot d\boldsymbol{B}_s - \frac{1}{2} \int_0^t \| \boldsymbol{\eta}_s \|^2 \, ds \right\},$$

i.e. L(t) is the solution to the stochastic differential equation

$$dL(t) = -L(t)\boldsymbol{\eta}_t \cdot d\boldsymbol{B}_t, \qquad L(0) = 1.$$

Assume (Novikov's condition) that

$$E\left\{ \exp\left(\frac{1}{2} \int_0^T \| \boldsymbol{\eta}_s \|^2 \, ds \right) \right\} < \infty.$$

Then

1. $L(T)$ is strictly positive and $E^P[L(T)] = 1$.
2. $L(T)$ and the associated density process $L(t) = E^P[L(T) \mid \mathcal{F}_t]$ generate an equivalent probability measure Q by the Radon–Nikodym derivative $\frac{dQ}{dP} = L(T)$.
3. $\tilde{\boldsymbol{B}}_t = \boldsymbol{B}_t + \int_0^t \boldsymbol{\eta}_s ds$ is a d-dimensional standard Brownian motion with respect to the measure Q.

Consider now the vector price process

$$d\tilde{\boldsymbol{S}}_t = \tilde{\boldsymbol{\mu}}_t dt + \tilde{\boldsymbol{\sigma}}_t \cdot d\boldsymbol{B}_t = (\tilde{\boldsymbol{\mu}}_t - \tilde{\boldsymbol{\sigma}}_t \cdot \boldsymbol{\eta}_t)dt + \tilde{\boldsymbol{\sigma}}_t \cdot d\tilde{\boldsymbol{B}}_t. \qquad (5.80)$$

If $\tilde{S}_t$ is going to be a Q-martingale, the η_t process must be chosen such that

$$\tilde{\mu}_t = \tilde{\sigma}_t \cdot \eta_t. \tag{5.81}$$

If $\tilde{\sigma}_t(\omega)$ has rank d for all $t \in [0,T]$ P-a.s., then $\tilde{\mu}_t = \tilde{\sigma}_t \cdot \eta_t$ has a solution. In the simple case, where $\tilde{\mu}_t$ and $\tilde{\sigma}_t$ are uniformly bounded and $\tilde{\sigma}_t$ is uniformly bounded away from zero, η_t will also be uniformly bounded and $E^P[L(T)] = 1$. This is also the case when $\tilde{\mu}_t$ and $\tilde{\sigma}_t$ are subject to a common factor of proportionality, the cancellation of which renders the solution η_t to (5.81) uniformly bounded.

Example 7 The famous Black–Scholes model has two assets: the bank account with a constant interest rate r, which is used as numéraire, and a single risky asset, whose price process is a geometric Brownian motion

$$dS_t = \mu S_t dt + \sigma S_t dW_t \tag{5.82}$$

$$S_t^0 = e^{rt} \tag{5.83}$$

$$d\tilde{S}_t = (\mu - r)\tilde{S}_t dt + \sigma \tilde{S}_t dW_t \tag{5.84}$$

$$d\tilde{S}_t = \sigma \tilde{S}_t d\left(W_t + \frac{\mu - r}{\sigma} t\right). \tag{5.85}$$

This is a case where the numéraire-modified drift term $(\mu - r)\tilde{S}_t$ and the numéraire-modified volatility term $\sigma \tilde{S}_t$ are subject to a common factor of proportionality, namely $\tilde{S}_t$. After cancellation, η_t from (5.81) becomes a constant. Hence, Girsanov's Theorem is directly applicable with $\eta_t = \frac{\mu - r}{\sigma}$ and the Radon–Nikodym derivative

$$L(T) = \exp\left\{-\frac{\mu - r}{\sigma} W_T - \frac{1}{2}\left(\frac{\mu - r}{\sigma}\right)^2 T\right\}.$$

Consider a European call option with exercise price X and expiration date T. The price of this asset at time t, $0 \le t \le T$ is denoted by $C(S_t, t)$. Contractually, at the expiration date T this asset pays its owner

$$C(S_T, T) \equiv \max\{0, S_T - X\}, \tag{5.86}$$

whereas the price at time 0 is given by

$$\begin{aligned} C(S_0, 0) &= E^P\left[L(T) \cdot e^{-rT} \cdot \max\{0, S_T - X\}\right] \\ &= S_0 \Phi(d_0 + \sigma\sqrt{T}) - X e^{-rT} \cdot \Phi(d_0), \end{aligned} \tag{5.87}$$

with

$$d_0 = \frac{\log(S_0/X) + \left(r - \frac{1}{2}\sigma^2\right) T}{\sigma\sqrt{T}},$$

where Φ is the cumulative distribution function for the standard normal distribution.

The price at time t is given as

$$C(S_t, t) = \frac{E^P\left[L(T) \cdot e^{-r(T-t)} \cdot \max\{0, S_T - X\} \mid \mathcal{F}_t\right]}{E^P\left[L(T) \mid \mathcal{F}_t\right]} =$$

$$S_t \Phi(d_t + \sigma\sqrt{T-t}) - X e^{-r(T-t)} \cdot \Phi(d_t), \tag{5.88}$$

with

$$d_t = \frac{\log(S_t/X) + \left(r - \frac{1}{2}\sigma^2\right)(T-t)}{\sigma\sqrt{T-t}}.$$

The self-financing trading strategy that makes this asset attainable is given by

$$\theta_S(t) = \Phi(d_t + \sigma\sqrt{T-t}) \qquad \theta_0(t) = -X e^{-r(T-t)} \Phi(d_t).$$

We will not pursue any further the necessary regularity conditions and the circumstances under which this insight holds in a broader setting. It is apparent that the existence of an equivalent martingale measure is closely related to the specification of the $\boldsymbol{S}$-processes and thereby the way that information is released. Recall that $\mathcal{F}_t = \sigma\left(\boldsymbol{S}_s, 0 \leq s \leq t\right)$. Hence, an important factor is the 'variety' of new information – as measured by rank$\{\boldsymbol{\sigma}_t\}$ – released by observing the development of prices in the market.

5.12 Complete market in continuous time

Definition 16 The market is said to be complete if any stochastic variable $X_T \in L^2(\Omega, \mathcal{F}_T, P)$ can be attained through a self-financing trading strategy:

$$X_T = \boldsymbol{\theta}_T \cdot \boldsymbol{S}_T = \boldsymbol{\theta}_0 \cdot \boldsymbol{S}_0 + \int_0^T \boldsymbol{\theta}_t \cdot d\boldsymbol{S}_t.$$

In order to obtain 'nice' results we will assume that $X_T \in L^2(P) \Leftrightarrow X_T \in L^2(Q)$. This will be the case, for example, whenever $\frac{dQ}{dP}$ is uniformly bounded from above as well as away from zero from below. The next theorem is the continuous-time counterpart to Theorem 10.

Theorem 15 Assume that $\boldsymbol{X}_T \in L^2(Q)$, where Q is an equivalent martingale measure. Then $\boldsymbol{X}_T$ is attainable if and only if there exists a predictable process $\{h_t\}$ such that

$$E^Q\left(\int_0^T \| h_t \cdot \tilde{\sigma}_t \|^2 \, dt\right) < \infty$$

$$\begin{aligned} E^Q\left(\tilde{X}_T \mid \mathcal{F}_t\right) &= E^Q\left(\tilde{X}_T\right) + \int_0^t h_s \cdot \tilde{\sigma}_s \cdot d\tilde{B}_s \\ &\quad \forall 0 \leq t \leq T. \end{aligned} \tag{5.89}$$

Proof Assume that X_T, and thereby $\tilde{X}_T$, is attainable:

$$\tilde{X}_T \equiv \theta_T \cdot \tilde{S}_T = \theta_0 \cdot \tilde{S}_0 + \int_0^T \theta_t \cdot d\tilde{S}_t.$$

Then

$$\begin{aligned} \tilde{X}_T &= \theta_0 \cdot \tilde{S}_0 + \int_0^T \left\{\theta_t \cdot \tilde{\mu}_t dt + \theta_t \cdot \tilde{\sigma}_t \cdot (d\tilde{B}_t - \eta_t dt)\right\} \\ &= \theta_0 \cdot \tilde{S}_0 + \int_0^T \theta_t \cdot \tilde{\sigma}_t \cdot d\tilde{B}_t. \end{aligned}$$

We conclude from here that $E^Q(\tilde{X}_T) = \theta_0 \cdot \tilde{S}_0$ and

$$E^Q\left(\tilde{X}_T \mid \mathcal{F}_t\right) = E^Q\left(\tilde{X}_T\right) + \int_0^t \theta_s \cdot \tilde{\sigma}_s \cdot d\tilde{B}_s.$$

Since $\{\theta_t\}$ is predictable the result follows.

To prove the reverse, let h_t be such that the conditions in (5.89) are fulfilled. We will then construct a self-financing trading strategy that generates $\tilde{X}_T$. Let

$$\theta_t^i = h_t^i \quad \text{for } i = 1, 2, \ldots, N, \tag{5.90}$$

$$\theta_t^0 = E^Q\left(\tilde{X}_T\right) + \int_0^t h_s \cdot \tilde{\sigma}_s \cdot d\tilde{B}_s - \sum_{i=1}^N \theta_t^i \tilde{S}_t^i. \tag{5.91}$$

Then

$$\theta_0 \cdot \tilde{S}_0 = E^Q\left[\tilde{X}_T\right], \tag{5.92}$$

$$\theta_t \cdot \tilde{S}_t = \theta_0 \cdot \tilde{S}_0 + \int_0^t \theta_s \cdot \tilde{\sigma}_s \cdot d\tilde{B}_s. \tag{5.93}$$

Hence, θ is a self-financing trading strategy, and from (5.89) with $\mathcal{F}_T$ we get that

$$\tilde{X}_T = \theta_0 \cdot \tilde{S}_0 + \int_0^T \theta_s \cdot \tilde{\sigma}_s \cdot d\tilde{B}_S. \tag{5.94}$$

Theorem 16 Assume that an equivalent martingale measure Q exists. Then the market is complete if rank$\{\tilde{\sigma}_t\} = d\,\forall t$.

Proof Let $\tilde{Y}_T$ be an arbitrary stochastic variable with finite variance, i.e. $\tilde{Y}_T \in L^2(Q)$ (=$L^2(P)$). From the martingale representation theorem we know that there exist predictable processes $\boldsymbol{h}_t \equiv (h_t^1, h_t^2, \ldots, h_t^d)$, such that $E^Q\left[\int_0^T \| \boldsymbol{h}_s \|^2 ds\right] < \infty$ and

$$E^Q\left[\tilde{Y}_T \mid \mathcal{F}_t\right] = E^Q\left[\tilde{Y}_T\right] + \int_0^t \boldsymbol{h}_s \cdot d\hat{\boldsymbol{B}}_s. \tag{5.95}$$

Whenever $\tilde{\sigma}_t$ has rank d, the equation $\boldsymbol{\theta}_t \cdot \tilde{\boldsymbol{\sigma}}_t = \boldsymbol{h}_t$ has a solution $(\theta_t^1, \theta_t^2, \ldots, \theta_t^N)$ for all t. Asset no. 0 is used to 'close the gap' through

$$\theta_t^0 = E^Q\left[\tilde{Y}_T \mid \mathcal{F}_t\right] - \sum_{i=1}^N \theta_t^i \tilde{S}_t^i$$

such that $\boldsymbol{\theta}_t = (\theta_t^0, \theta_t^1, \ldots, \theta_t^N)$ becomes a self-financing trading strategy. By construction, $\boldsymbol{\theta}_t$ generates $\tilde{Y}_T$.

We will finish the discussion of continuous-time models with the continuous-time analogue to the splitting index in discrete time. Within the scope of this paper, we do not attempt to go into thorough proofs, but limit ourselves to a presentation of the main results.

By $\mathcal{M}_P^2$ we denote the space of square integrable martingales on $(\Omega, \mathcal{F}, P)$ with value 0 at $t = 0$. The spaces $L^2(P)$ and $\mathcal{M}_P^2$ are in one-to-one correspondence with each other through the relation $X(t) = E^P[x \mid \mathcal{F}_t]$, where $x \in L^2(P)$ and $X \in \mathcal{M}_P^2$. In this set-up it can be shown (see the references and examples given in Duffie and Huang, 1985) that there exists a least integer K with the property that there exist mutually orthogonal martingales $\{m^1, \ldots, m^K\}$ such that for any $X \in \mathcal{M}_P^2$ there exists a predictable vector process $\eta_t = (\eta_t^1, \eta_t^2, \ldots, \eta_t^K)$ such that

$$X_t = \int_0^t \eta_s dm_s.$$

Such a minimal set of mutually orthogonal martingales is denoted as an orthogonal '2-basis' – with reference to the number 2 in L^2 – for

$X \in \mathcal{M}_P^2$. K is said to be the **martingale multiplicity** of $\mathcal{M}_P^2$.

The equivalent concept in discrete time, the dimension of the martingale basis, was developed in section 5.9. We found that this dimension was related to the splitting function of the filtration, and in Theorem 9 it was stated that if $\nu(t,\omega) = p > 1 \ \forall (t,\omega) \in \mathcal{T} \times \Omega$, then $p-1$ was the dimension of any martingale basis. K is the continuous-time equivalent of the number $p-1$ in discrete time.

Furthermore, we concluded in section 5.8 that a necessary requirement in order to ensure completeness was that the number of assets $N+1$ should satisfy $N+1 \geq \max\{\nu(t,\omega) \mid (t,\omega) \in \mathcal{T} \times \Omega\}$. Without proof we state the equivalent theorem in the continuous-time economy in the situation where the spaces $L_P^2(m)$ and $L_Q^2(m)$ coincide. This is guaranteed to be the case, whenever $\frac{dQ}{dP}$ is uniformly absolutely continuous.

Theorem 17 The minimal number of assets necessary in order to render the market complete is $K+1$.

5.13 Equivalent martingale measures and absence of 'approximate arbitrage'

The set of stochastic variables that are attainable by using a self-financing trading strategy is a linear subspace, M_0, of $L^2(P)$. Absence of arbitrage means that the pricing functional must be a strictly increasing linear functional $\pi : M_0 \to R$ such that $\pi(Z_T) = \theta_0 \cdot Z_0 \ \forall Z_T \in M$.

In an infinite state space there is no guarantee that the pricing functional π is continuous and that the subspace M_0 is closed. Hence, the previously applied extension technique, generating an equivalent martingale measure by extending π to a strictly increasing and continuous linear functional on all of $L^2(P)$, is not guaranteed to work. If such an extension was possible, the Riesz representation theorem and the choice of a feasible numéraire would immediately give the recipe for the construction of the Radon–Nikodym derivative for an equivalent martingale measure, under which the relative prices of all attainable assets would be martingales.

Rather than assuming that M_0 is closed and π is continuous, we will state a recent result related to the concept of 'approximate no-arbitrage' in the $L^2(P)$ setting.

Definition 17 An approximate arbitrage opportunity is a sequence $\{z_n\}$ in M_0, with $\pi(z_n) \leq 0 \ \forall n$, such that $z_n \to z$, $z \geq 0$, $P\{z > 0\} > 0$

and $\lim_n \pi(z_n) \leq 0$.

Theorem 18 If no approximate arbitrage opportunities exist, and the bank account – with a bounded short-term interest rate – serves as the numéraire, then an equivalent martingale measure exists, under which the relative price processes are martingales.

The procedure for proving this theorem is first to extend π from M_0 to the closure $\overline{M_0}$ in a continuous and strictly increasing manner. Given this extension the Riesz representation theorem takes care of the rest.

To obtain the entire foundation for the very brief presentation in this section, a substantial amount of technical machinery is needed. The relevant references are Delbaen and Schachermayer (1994), Clark (1993) and Chapter 6 in Duffie (1992).

5.14 Concluding remarks

The central topics in this paper,

- the equivalence between 'no arbitrage' and the existence of an equivalent martingale measure
- the theorems concerning market completeness and unique pricing of contingent claims

have been focal topics in many monographs as well as journal articles, since the seminal publications by Harrison and Kreps (1979) and Harrison and Pliska (1981).

As for books, the monographs by Ingersoll (1987), Dothan (1990), Duffie (1992), Lamberton and Lapeyre (1992) and Dana and Jeanblanc-Picqué (1994) present the main theorems in various settings and in various degrees of mathematical complexity. During the preparation of this article, we have found the lecture notes by Björk (1994) very useful.

The connection between equivalent martingale measures and the absence of arbitrage and the difference between models in continuous and discrete time are both subjects that are extensively discussed in the literature. At the risk of omitting some important references, we recommend that the reader consult the following relevant articles: Harrison and Pliska (1983), Duffie and Huang (1985), Huang (1985), Taqqu and Willinger (1987), Dybvig and Huang (1988), Dalang, Morton and Willinger (1990), Stricker (1990), Back and Pliska (1991), Clark (1993), Delbaen and Schachermayer (1994) and Kabanov and Kramkov (1994). The introduction to the paper by Delbaen and Schachermayer (1994), which was published during the preparation of this article, has a short survey of the historical development of the equivalence between 'no arbitrage' and the absence of arbitrage.

Market completeness and the uniqueness/non-uniqueness of the equivalent martingale measure and of the pricing functional is the central theme in the papers by Müller (1989) and Jarrow and Madan (1991). Unique pricing in an incomplete market has not been discussed in this paper for the simple reason that an incomplete market gives rise to multiple martingale measures and non-unique pricing functionals. It is not possible to choose between this multiplicity of martingale measures without introducing a preference ordering or an *ad hoc* criterion for selection. Introducing a preference ordering has a long history, with Rubinstein (1976) as a seminal paper along those lines. Some efforts have been devoted in the literature to pricing in incomplete markets by applying an *ad hoc* criterion of risk minimization. Contributions along these lines can be found in e.g. Föllmer and Sondermann (1986) and in Schweitzer (1988; 1991; 1994).

Acknowledgements

Financial support from the Danish Social Science Research Council and the Danish Natural Science Research Council is gratefully acknowledged. We are grateful for comments and suggestions from Ole E. Barndorff-Nielsen, Asbjørn Hansen, David Lando, Ralf Korn, Peter N.D. Møller and Carsten Sørensen. Needless to say, they are in no way responsible either for the content of the paper or for remaining errors.

References

Back, K. and Pliska, S.R. (1991) On the fundamental theorem of asset pricing with an infinite state space. *Journal of Mathematical Economics*, **20**, 1–18.

Björk, T. (1994) *Stokastisk kalkyl och kapitalmarknadsteori*. Lecture notes, KTH, Stockholm.

Black, F. and Scholes, M. (1973) The pricing of options and corporate liabilities. *Journal of Political Economy*, **81**, 637–659.

Clark, S.A. (1993) The valuation problem in arbitrage price theory. *Journal of Mathematical Economics*, **22**, 463–478.

Cox, J.C., Ross, S.A. and Rubinstein, M. (1979) Option pricing: a simplified approach. *Journal of Financial Economics*, **7**, 229–263.

Dalang, R.C., Morton, A.J. and Willinger, W. (1990) Equivalent martingale measures and no-arbitrage in stochastic securities market models. *Stochastics and Stochastic Reports*, **29**, 185–201.

Dana, R.-A. and Jeanblanc-Picqué, M. (1994) *Marchés Financiers en Temps Continu*, Economica, Paris.

Delbaen, F. and Schachermayer, W. (1994) A general version of the fundamental theorem of asset pricing. *Mathematische Annalen*, **300**, 463–520.

Dothan, M.U. (1990) *Prices in Financial Markets*. Oxford University Press, Oxford.

Duffie, D. (1992) *Dynamic Asset Pricing Theory*, Princeton University Press, Princeton, NJ.

Duffie, D. and Huang, C.-F. (1985) Implementing Arrow–Debreu equilibria by continuous trading of few long-lived securities. *Econometrica*, **53**, 1337–1356.

Dybvig, P. and Huang, C.-F. (1988) Nonnegative wealth, absence of arbitrage, and feasible consumption plans. *Review of Financial Studies*, **1**, 377–402.

Föllmer, H. and Sondermann, D. (1986) Hedging of non-redundant contingent claims. In W. Hildenbrand and A. Mas-Colell (eds), *Contributions to Mathematical Economics* North-Holland, Amsterdam.

Harrison, M.J. and Kreps, D.M. (1979) Martingales and arbitrage in multiperiod securities markets. *Journal of Economic Theory*, **20**, 381–408.

Harrison, M.J. and Pliska, S.R. (1981) Martingales and stochastic integrals in the theory of continuous trading. *Stochastic Processes and their Applications*, **11**, 215–260.

Harrison, M.J. and Pliska, S.R. (1983) A stochastic calculus model of continuous trading: complete markets. *Stochastic Processes and their Applications*, **15**, 313–316.

Huang, C.-F. (1985) Information structures and viable price systems. *Journal of Mathematical Economics*, **14**, 215–240.

Ingersoll, J.E. (1987) *Theory of Financial Decision Making*. Rowman & Littlefield, Savage, MD.

Jarrow, R.A. and Madan, D.P. (1991) A characterization of complete security markets on a Brownian filtration. *Mathematical Finance*, **1**, 31–43.

Jensen, B.A. and Nielsen, J.A. (1992) The structure of binomial lattice models for bonds. Working Paper WP 92–17: Institute of

Finance, Copenhagen Business School. To appear in A.A. Novikov (ed.), *Statistics and Control of Stochastic Processes.* TVP, Science Publishers, Moscow.

Kabanov, Y.M. and Kramkov, D.O. (1994) No-arbitrage and equivalent martingale measures: an elementary proof of the Harrison-Pliska theorem. Preprint: Central Economics and Mathematics Institute, Moscow.

Lamberton, D. and Lapeyre, B. (1992) *Introduction au calcul stochastique appliqué à la finance,* ELLIPSES, Paris.

Modigliani, F. and Miller, M.H. (1958) The cost of capital, corporation finance, and the theory of investment. *American Economic Review,* **48**, 261–297.

Müller, S.M. (1989) On complete securities markets and the martingale property of security prices. *Economic Letters*, **31**, 37–41.

Rockafellar, R.T. (1970) *Convex Analysis.* Princeton University Press, Princeton, NJ.

Rubinstein, M. (1976) The valuation of uncertain income streams and the pricing of options. *Bell Journal of Economics*, **7**, 407–425.

Schweitzer, M. (1988) *Hedging of Options in a General Semimartingale Model,* Ph.D. dissertation, ETHZ, Zürich.

Schweitzer, M. (1991) Option hedging for semimartingales, *Stochastic Processes and their Applications*, **37**, 339–363.

Schweitzer, M. (1994) Risk-minimizing hedging strategies under restricted information. *Mathematical Finance*, **4**, 327–342.

Stricker, C. (1990) Arbitrage et lois de martingale. *Annales de l'Institut Henri Poincaré – Probabilités et Statistiques*, **26**, 451–460.

Taqqu, M.S. and Willinger, W. (1987) The analysis of finite security markets using martingales. *Advances in Applied Probability*, **19**, 1–25.

Vasicek, O. (1977) An equilibrium characterization of the term structure. *Journal of Financial Economics*, **5**, 177–188.

Printed in the United States
by Baker & Taylor Publisher Services